CARCERAL CONTEXTS:
READINGS IN CONTROL

EDITED BY
KEVIN R.E. MCCORMICK

Canadian Scholars' Press Inc.   Toronto   1994

Carceral Contexts: Readings in Control

First published in 1994 by
Canadian Scholars' Press Inc.
180 Bloor St. W., Ste. 402,
Toronto, Ontario   M5S 2V6

Canadian Cataloguing in Publication Data

Main entry under title:
Carceral Contexts: Readings in Control
Includes bibliographical references.

ISBN   1-55130-007-9

1. Corrections. I. McCormick, Kevin R.E., 1965-

HV8705.C37 1994   365   C94-930100-0

Printed and bound in Canada

Dedication

To Renee, my wife

whose intelligence, compassion and love

fill my life with meaning and purpose.

Acknowledgements

This work would have never been possible without the support and guidance of a number of individuals, to whom I remain truly indebted.

Firstly, I would like to thank the contributors to this collective enterprise, who, representing a myriad of perspectives, must be commended for their outstanding work. Not only do their works represent critical research on the nature of control and power in the criminal justice system, but also reflect their ongoing commitment to education at its most critical level.

I am also very appreciative of the support and direction given me by a number of students at both York University and Atkinson College. Their enthusiasm and passion for education serve as a constant reminder of what a true privilege it is to be a teacher.

Further, I am indebted to a number of colleagues for their encouragement, insights and counsel. These colleagues are Paul Bessler, Helcinto Menezes, Florence Pirruza, Dragan Spasojevic and Robert Stevens, whose support throughout this career is very much appreciated.

I would also like to thank those individuals who, while not contributing directly to this document, nevertheless have aided the editor in numerous ways. To the support staff at Atkinson College, Department of Sociology, Rabia Sallie and Debbie Keltz, thank you for all the assistance that you have given me throughout the years. Your concern for quality education is very much appreciated. Further, to Erica, Holly, Sharon, Shirley, Suzanne, Lidia, Lorraine and Pauline in the Faculty of Graduate Studies, thank you for your assistance, without which this project would have never been possible.

Further, I would also like to extend my appreciation to Canadian Scholars' Press Inc., especially Jack Wayne, Brad Lambertus and Pamela Hamilton. Throughout this enterprise their commitment to quality and scholarship was very much appreciated by the editor.

Finally, I would like to extend my thanks to my family for their unconditional support. To the new members of my family, Elizabeth, Pamela, Raymond and

Derrick, for welcoming me into their lives. Especially, I would like to express my appreciation and love to my Mother, Doris, for being the one I could always count on and to my Father, Kenneth, for his guidance and unconditional support.

This project was supported in part by the Social Science and Humanities Research Council of Canada. Opinions of the authors and the editor do not necessarily reflect those of the council .

Editor's Biography

Kevin R.E. McCormick is a doctoral candidate in the Department of Sociology at York University and a Fellow of Bethune College. His work is directed at critically examining the corrigibility of traditional penological theory and research to address the interrelationship between the offender and the physical, technological and ideological barriers in which he/she is imprisoned. His research challenges and expands the empirical and conceptual parameters of contemporary criminological discourse. His published works, as demonstrated in *Canadian Penology* and *Understanding Policing,* co-edited with Livy A. Visano, investigate and examine the social constructions of the juridic actor within countless social and institutional contexts.

Table of Contents

Foreword

Claudio Duran

It is with great honour that I introduce *Carceral Contexts*, a book whose conception is consistent with some of my basic academic principles. I have always believed that teaching and research, including publication, are two sides of the same coin. The pedagogical act must, therefore, involve processes that keep that close relationship alive. Students learn the complexity of the relationship with the assistance of the teacher and at the same time, as Martin Heidegger suggests, in all true teaching the teacher learns as well. A teacher should address the pedagogical experience as a sublime and natural moment of profound sharing with the students: a sharing that will bring further awareness of what knowledge is about.

In that sense, the importance and meaning of the work produced by a student during a course is more than just a grade. Grades are obviously important since they are a quantitative and qualitative reflection of the learning process. When grading is the result of serious marking and commenting by the teacher it becomes a true pedagogical experience. However, many times students produce work that goes beyond the requirements of an assignment. Perhaps this is more likely to happen at the higher levels of undergraduate education and in graduate school, but, in any event, when it happens the teachers feel that their efforts have been especially rewarded.

From my own personal experience, I can report that one of my own undergraduate students, who is now a graduate student in philosophy, wrote an extraordinary paper on the psychoanalytic theory of bi-logic. He was introduced to my colleagues in England and Chile and they were as impressed with the paper as I was. He presented the paper in England and it was included in a book recently published in Chile. This was a book in which an article of my own also appeared. This has been a most rewarding experience.

I came to Canada with my family as a political refugee. We arrived in Montreal on October 7, 1973, only a month after the military coup in Chile. It was Thanksgiving weekend. A few days later we came to Toronto and, by

November, I was teaching at York University. I was offered a position, thanks to a number of colleagues, from the Department of Philosophy at Atkinson College. At first, I also taught in the Department of Visual Arts and in the division of Social Sciences in the Faculty of Arts.

York University impressed me from the start because it was new and innovative and because it was committed to teaching an interdisciplinary approach. Moreover the academic environment was challenging and creative and based on democratic and participatory ideals. These ideals manifested themselves in small classes, the encouragement of discussion and student participation. I felt very comfortable here because I could work within the general pedagogical framework that I had been familiar with in Chile where I had taught since 1967.

Atkinson College and my Department provided me with a 'home' in the truest sense of the word. I have always been fascinated by students because they respect you if you respect them and they teach you to be a more humble teacher. Their motivation for learning is extremely high; this in turn inspires their teachers. A sense of friendship develops and contributes very much to the enhancement of learning. Of course, this learning is only possible when the teacher participates with the students in the process of learning. In philosophy, for example, the teacher encourages the students to read and analyze original philosophical texts on their own and most of the class time is devoted to the exchange of views about those texts. Lectures are organized in accordance with a participatory approach.

This is what is so impressive about *Carceral Contexts*. Kevin McCormick has gathered a representative and well-rounded collection of papers from students and teachers, from service organizations and from the academic community. These papers will make a significant contribution to our under-standing of criminal justice. It would have been to our collective detriment as a community if these works had gone unpublished. Moreover, by publishing these papers Kevin McCormick is enhancing that view of the pedagogical experience that I discussed earlier. I applaud his participatory approach.

In conclusion, I must congratulate the editor for his great inspiration in the conception of this book, but also the contributors whose papers you will have the privilege of reading.

Professor Claudio Duran
Atkinson College, York University
1993 C.A.S.E. Canadian Professor of the Year

Preface

Carceral Contexts and Political Contests: Challenging the Boundaries

The Canadian penal system is currently experiencing a dramatic period of both change and public inquiry. Community indignation over issues of parole, probation, prisonization and rehabilitation has resulted in the enactment of numerous policy and program initiatives by various sectors of the criminal justice system, including corrections and policing. While meticulous care has been taken by those in power to establish a philosophy that conveniently redefines traditional penal enterprises, the fundamentally problematic nature of corrections remains systematically ignored. This failure to critically address the needs of those empowered with the responsibility of both interpreting and delivering correctional mandates impacts dramatically on the entire penal system, resulting in wide disparities in the delivery of services and concomitant interpretations of policy changes.

This collection of articles examines the construction of social control as constituting a carceral context, a convenient mythology that perpetuates, conceals and justifies the coercive nature of the criminal justice system. Specifically, this work generates a theoretical, methodological and pedagogical foundation upon which to facilitate a greater comprehension of the systemic racism, sexism, ageism, homophobia and ablism inherent in various correctional and policed environments. Conceptually and substantively, strategies will be articulated which seek to overcome the above-noted impediments and recommendations will be offered which impact directly upon future policy and program initiatives.

Through deliberate design, this collection of articles merges distinct disciplines and diverse backgrounds. Contributors representing the disciplines of sociology, penology, criminology, social work, gerontology and law will establish a dynamic paradigm to interrogate the problematic practices which produce and perpetuate the misogynist ideologies pervading the justice system. Further, this work melds together scholars from various backgrounds, including university and college professors, community advocates and students, each bringing to the text rich experiences within which their discussions of the carceral are situated. This diverse collective of leading researchers was charged with the responsibility of addressing constructions and contextualizations of the carceral in a fashion that reached various audiences, including undergraduates, graduate students, social workers and justice personnel.

Those who read this collection of articles will be empowered to challenge the normative assumptions within which definitions of power and control are contextualized. Reaching past the supposedly 'rational' and 'just' pronouncements of government policies and programs, accordingly it will be argued that control transcends convenient reflections of safety and security. Control is conceptualized as contests premised on issues and practices of power and privilege. By systematically questioning the common sense notions upon which the criminal justice system was founded, students reading this text will be able to critically challenge the institutional practices which perpetuate these ideologies and develop individual and collective strategies for social change.

K. M.

Official Rhetoric and the Disposal of Reality

I - Introduction

The recent flurry of government justice bills suggests that once again with federal elections around the corner, 'law and order' reigns supreme on Parliament Hill. Indeed, cynicism about legislation regarding sentencing reform, child pornography, stalking, sexual predators and drug use appears justified in view of government pandering to public fears about crime. The scope of these issues makes it doubtful that such proposals could receive proper scrutiny before they became law, especially when discussion about them is mired in contradiction over the competing wisdoms of rehabilitative, deterrent and retributive philosophies. While the declared purpose of all these reforms is to create a 'just and peaceful society,' the strenuous insistence on a 'safe society' promises new impositions of punishment. Fiscal considerations also intrude, so visionary alternatives to prison are left on the drawing-board — even new prison construction subject to instant cutbacks. As well, the clarion call to 'streamlining of objectives' does little to inspire correctional personnel who know from past applications the regressive effects of austerity. It would seem that the credibility of the entire criminal justice system is diminished day-by-day as its isolation from local communities becomes more apparent, and what passes for 'community involvement' sweeps a precarious arc from bureaucratic indifference to vigilantism.

Still, 'involvement' is the watchword, as government functionaries strive to implicate 'community' in the process of justice reform. If it is to be more than a mere shibboleth, however, 'community' must translate into representation, input and impact in the shaping of criminal law and its enforcement. It would be ironic, for example, if the avowed aim of 'successfully reintegrating the offender into the community' were carried out in ways that ignored the will and capacities of a community to fulfill that purpose. But this is the likely effect of rushing government anti-crime bills through Parliament, a consequence known to all who partake in such politically expedient charades.

May we, however, look for different outcomes amongst those government institutions ostensibly devoted to sober and reflective examination of these very government proposals? Do House Committee 'hearings,' for example, truly engage and actively 'hear' the community voices relevant to consideration of a given government bill? Do members of these committees display reliable neutrality and fairness before the array of 'witnesses' who comprise the varied interests of 'community,' and whose representations are solicited presumably because they enhance the rationality of government decision-making? Or do these hearing panels merely legitimate preconceived government objectives? If *abbreviated* hearings suggest disregard for the myriad voices of community, does that mean that *lengthy* and *ceremonious* procedures imply unbiased deliberations?

Since 'hearings' conducted under the aegis of Standing Committees of the House or Senate are ordinarily held in high esteem by those who elect to participate in them, the above questions are important as hypocrisy is harder to unmask but no less prevalent in high places. Specifically, I wish to examine proceedings of the House Standing Committee on Justice and the Solicitor General in its hearings held on March 9th and 10th, 1992 in Vancouver, B.C. regarding Bill C-36, the *Corrections and Conditional Release Act*, an *Act* respecting corrections and the conditional release and detention of offenders and to establish the office of Correctional Investigator. As a representative of the British Columbia Civil Liberties Association, I presented a brief to the Committee and I remained to observe other presentations over the two-day hearing. The bill, which had passed second reading prior to the hearing, was a calculated effort by the Progressive Conservative government to respond to public perceptions of increasing crime coupled with state leniency toward offenders. The provisions of the bill were primarily aimed at restricting the availability of parole for violent offenders, toughening penalties against drug users and sex offenders, and balancing these punitive features with accelerated parole eligibility for non-violent offenders.[1]

In evaluating whether this particular hearing fulfilled its avowed mandate to consider the views of the members of 'community' who came before it, I will invoke some of the ideas of Jurgen Habermas, a contemporary critical theorist — first, to specify the criteria for optimum dialogue in such deliberations, and second, to examine some crucial exchanges in the hearings which I participated in and observed, in order to determine whether those criteria prevailed.

II - Rational Discourse

Habermas understands society as evolving through a process of structural differentiation causing integrative tensions that peak in "legitimation" and "motivational" crises.[2] He believes that these crises can be resolved if the processes that create unifying cultural symbols are given equal weight with the processes of material production. Industrial society has so far failed to achieve this degree of integration; instead, "system-processes" in the economic and political realms have colonized the "lifeworld" processes at the interactive/cultural level.[3] Impersonal steering mechanisms of money and power — the "delinguistified media" — control the interpretation of everyday existence, resulting in "power-distorted communication" that subordinates the social creation of meaning to actions based on predetermined ends. Habermas's emancipatory formula is to decrease instrumentalization and increase meaningful symbolic communication, thus restoring the balance between lifeworld and system processes.

"Communicative competency," therefore, depends on the realization (as nearly as possible) of the "ideal speech situation,"[4] which bears four recognizable characteristics:

(1) **"truth"** - the statements adequately portray realities in relation to the objective world;

(2) **"truthfulness"** - the speaker is sincere and authentic in the expression of intentions and feelings;

(3) **"comprehensibility"** - the speaker uses language adequately enough to be understood; and

(4) **"understandability"** - statements are appropriate or legitimate in a given situation in relation to shared norms and values.

While these "validity claims" are frequently violated in actual communication between speakers in modern industrial societies, Habermas believes that social interaction will disintegrate to the extent that these criteria lose applicability. The antidote to their irrelevance is the removal of institutional constraints that hamper speakers from challenging or accepting validity claims, thus enabling them to examine the sources of pseudo-communication. This cannot be done when claims are settled by recourse to power and authority. For validity claims to be mediated, speakers must have symmetrical chances to speak, the right to question whatever assumptions govern a situation, and the freedom to reject instrumental motivations. By ensuring these conditions for "rational discourse," Habermas anticipates a resurrection of the "public sphere," one that will reconnect system processes and lifeworld, infusing new meaning and commitment in social life and creating a "discursively redeemable" or just society. Restoration of the public sphere is signalled, therefore, by genuine debate and argumentation which replaces power and authority in political decision-making. When such desired forms of argumentation pervade cultural institutions (e.g., the legal system), they serve as mechanisms for system-integration by fostering communicatively mediated realities. By this standard, one might expect that legislative hearings would also constitute sites of mediation that exhibit the "institutional unboundedness" conducive to realization of the ideal speech situation.

In considering Habermas' s seminal concepts,[5] the question that arises for me is whether the House Standing Committee hearings which I attended on Bill C-36 emulated the characteristics of "rational discourse," or whether committee members were guided by the kinds of "strategic motivations" that Habermas associates with power-distorted communication. Were these hearings a "reflective medium" for settling problematic validity claims in a quest for mutual understanding, or were they a political artifice intended to mimic the outward features of "rational discourse" in order to pre-empt criticism and dissent? The answer to this question may further an understanding of the hegemonic power of the state in precisely those areas where it professes to relinquish some of its strength. The question is particularly intriguing since the government's need to provide manifestly democratic forums for 'public input' — and thereby fulfill a necessary legitimation function — runs counter to the tactical contingency of approximately 40 Department Ministers vying for limited legislative table-time, which, in turn, suggests that once Ministers clear proposed bills with Cabinet, they are unlikely to relent on their main features; consequently, hearings would be very much after the fact.[6]

III - Standing Committee Hearings

Bill C-36 (*Corrections and Conditional Release Act*) was tabled for first reading on October 8th, 1991. It was given second reading on November 4th and 5th, and then referred to the Standing Committee on Justice and Solicitor General for public hearings. Hearings were held in various cities across Canada during February and March of 1992. The Standing Committee was composed of five Conservative MPs (including the Chair, Bob Horner), two Liberal MPs, and one MP representing the New Democratic Party.[7] Most of the eight members were present during most of the hearing sessions, although three of the Committee members were substituting for original appointees.

The Vancouver hearings were held in the Waddington Room of the Hotel Vancouver. Participants were seated around a large rectangular table, open at the center, with the Chair and presenters at opposite ends. Government and opposition MPs sat across from one another on the longer sides. Microphones were placed on the tables before participants, and high-tech recording equipment was stored at the far end, behind the Chair. Government aides and journalists were seated inside a narrow booth that ran along the entrance wall. Ample seating for prospective speakers and casual attenders was available directly behind the speakers' end of the table. The room was elegant but not forbidding. Light refreshments were continuously available.

In addressing the Habermasian criteria for "rational discourse" — truth, truthfulness, comprehensibility, and understandability — I will focus on three of the briefs presented on the first day of the hearings: my own, a panel of prisoner advocates and a delegation from the B.C. Criminal Justice Association. In demonstrating the criterion of "understandability," I will shift venue from the hearings to the House discussion of the Bill at the time of third reading. In all instances, I will refer to portions of the official transcripts for documentation of the relevant discussions, recognizing that these provide only anecdotal, if revealing, insights into the performative and dominant logic of the hearing occasion.

a) "Truth"

In my formal presentation on May 9th to the Standing Committee, I questioned the rationale for lumping offenders convicted of 'violent' offenses with drug offenders, stressing that there was no public outcry for addressing the 'drug problem' in this way. Nevertheless, as with the previous witness that morning, the Committee questioning quickly turned to the issue of drug offenses, the alleged link between drugs and violence, and the need to prevent drugs from coming into the prisons. Under questioning by the NDP Member, I tried to explain why it would be difficult, perhaps impossible, to eliminate drug use from the penitentiary setting, a line of reasoning that seemed, at best, to evoke skeptical reactions from Members.

Mr. Blackburn: Obviously this is a very serious matter and something has to be done about it. The general public are confounded. We build prisons that virtually prevent the person from escaping, from getting out, but we can't stop the drugs from going in. It's illogical. It doesn't make any sense to me. Why is this going on? I know you're not responsible for our correctional services and for the administration of our prisons or correctional institutions, or whatever they're called.

Prof. Ratner: Even the idea of zero tolerance of drugs in penitentiaries may be unrealistic, frankly speaking.

Mr. Blackburn: Why?

Prof. Ratner: Because the penitentiary regime is such that drugs provide one of the few solaces, escapes, ways of coping with long-term imprisonment, which is not to say that the regime itself...

Mr. Blackburn: Excuse me. Couldn't that kind of medication be administered legally by medical staff, under controlled circumstances? We do that outside of prisons if people are suffering from depression or a manic state of mind, or if they are schizophrenic, or whatever it happens to be. If they do have an emotional disorder, sometimes physicians will prescribe certain medications. I see nothing wrong with that on the inside. That's a basic human right.

Prof. Ratner: Yes, if it's a medicinal problem, certainly, but if it's a recreational matter, no.

Mr. Blackburn: I'm not suggesting that.

Prof. Ratner: But I'm saying that for many of the inmates who indulge, it is a recreational matter. I'm not condoning it, but that is the impetus, and it's extremely difficult to eliminate totally.[8]

The Committee's apparent preoccupation with hammering home the importance of addressing the 'drug problem,' despite the remonstrations of expert or knowledgeable witnesses, was revealed in their questioning of the delegation that followed my segment. A highly articulate ex-inmate on day parole (he had spent ten years in penitentiaries and was now a member of the B.C. Prisoners' Rights' Group) made some disturbingly frank statements, beginning with his expression of regret "that the hearing will not be able to go to very many of our Canadian federal prisons, because we are talking about prisoners."[9] After some bullying questions meant to uphold the virtues of punishment, the NDP Member pursued a line of questioning, the answers to which left the Committee aghast.

Mr. Blackburn: Is it possible, Mr. Gastonguay, that because life was so terrible for you in prison you have decided that you didn't want to go back?

Mr. Gastonguay: Well, life wasn't terrible.

Mr. Blackburn: It wasn't?

Mr. Gastonguay: No, I used to have a personal computer. I used to have a colour TV, a radio. I used to do my own thing.

Mr. Blackburn: Were you in maximum or medium?

Mr. Gastonguay: I used to smoke every night — no big deal.

Mr. Blackburn: You smoked dope every night?

Mr. Gastonguay: Yes... I was reading Mr. Ingstrup's [then Commissioner of Corrections] testimony to this committee about drugs. He says they are being used occasionally.

Mr. Blackburn: They are being used all the time?

Mr. Gastonguay: All the time.

Mr. Blackburn: By a majority of the inmate population?

Mr. Gastonguay: Yes. And people become addicts in jail.

Mr. Blackburn: Did you?

Mr. Gastonguay: I didn't become...I got worse, yes, because I used to smoke every day. I am out now and I don't touch it.

Mr. Blackburn: You don't?

Mr. Gastonguay: I don't need to.[10]

The undisguised truthfulness of the witness was almost more than the Members could bear. One of them, apparently recalling my previously delivered comments about the prevalence of recreational drug use within prison grounds, twisted his head in my direction (I was now in the public seating area), unable to suppress an embarrassed grin. Interestingly, the next Member to question the witness, a Conservative from Quebec, elicited details on the witness's biography, specifically dwelling on his middle-class background and educational attainments in order to debunk his credentials as an 'average prisoner.' This defensive reaction to the witness's candor seemed designed to sustain the Members' imagined solution to the prison drug problem, i.e. the total eradication of non-medicinal drug use. Clearly, if recreational drug use in prisons actually contributed to institutional order, then its use could not be prohibited on the grounds that it was responsible for violence and disorder. The empirical 'truth' of drug use (assuming that the witness was not fibbing) was not assimilable by committee Members since it jeopardized a preconceived assessment linked to a virtually non-negotiable feature of the bill. Thus, the ability of Members to hear the 'truth' depended on whether it resonated with their contextual 'realityframe,' which, in this instance, could be equated with a set of political commitments that were Ministerially ordained. To the extent that such commitments required the exclusion of certain facts from consideration, 'rational discourse' was conducted within palpable if unofficially declared limits.

b) "Comprehensibility"

Whether witnesses and Members use language adequately enough to be understood also depends upon their respective needs and priorities as interpreters and speakers. This was confirmed by testimony of three members of the British Columbia Criminal Justice Association (BCCJA).

In its brief to the Committee, the BCCJA recommended against the toughening of penalties and urged that the proposal to classify drug offenses with "violent" offenses be abandoned. Their criticism of the government's proposals for sanctioning drug offenders was underscored by a hypothetical circumstance which drew disbelieving responses from Members.

> ...drug offenders will symbolically be sacrificed as an intellectual gesture to the war on drugs. Drug offenders have one of the highest success rates of any group of federal offenders, yet they will now be specifically targeted for detention if they are considered likely to commit a serious drug offence. This bill's definition of a serious drug offence would include sharing a marijuana with a friend...[11]

Since this criticism cast doubt on the justificatory centrepiece of Bill C-36, it proved most galling to the Committee. The response by one of the Liberal members illustrates the degree of pique.

Mr. Wappel:...before I do too much complimenting I want to take you severely to task for, on page 6 of your report, the paragraph about drug offences. I'm flabbergasted that the statement is in there that "This bill's definition of a serious drug offence would include sharing a marijuana cigarette with a friend." If it is in there, I agree with you that it's ridiculous. I can't believe it's in there. I can't find it in there. I do not believe the definition of a serious drug offence would include sharing a marijuana cigarette, I'd like you to show me where it is... It caused the credibility of your paper to suffer somewhat... If anyone would say to me, under any reasonable Charter argument, that a serious drug offence is sharing a joint, we'd better leave the country, because there's something wrong with the justice system.[12]

The author of the BCCJA submission responded to the Member's incredulity by explaining the possibility of such outcomes in terms of "unintended consequences" and "administrative repercussions" — i.e. the organizational processing of plainly innocuous acts such as "sharing a joint" leading to serious criminal charges. This quite plausible response made little impression on the Committee as both government and oppositional party Members continued to allude to the delegation's "loss of credibility" in "stretching the argument" via the joint-sharing vignette. After some discussion of other matters, one of the members of the delegation returned to the "credibility" issue, adding supportive information which again fell on deaf ears.

Ms. Hobbs-Birnie: Just before I conclude, Mr. Wappel, I would like to make a comment about the credibility of the report that is reflected in this comment about marijuana. When we sit in this room here and talk of legislation and how it will flow and be applied, in my opinion it is necessary to realize that the way it will be applied is not the way we imagine it will be applied, because once it reaches the prison level it goes into a different world, a very paranoid world, a highly punitive world. It is a world where indeed it is possible for someone to be severely punished by years for smoking a marijuana joint. During the course of my years on the board, I could give scores of incidences where something like that happened, where a man would be returned to prison and have to serve 'the remainder of his sentence for having smoked marijuana or having had a can of beer. I have sent five guys up for five years for that.

Mr. Blackburn: Five years?

Ms. Hobbs-Birnie: Whatever was remaining on their sentence. They would be hauled in, their parole would be revoked, and they would be back in.[13]

My point in drawing attention to the verbal struggles between the Committee and the delegation over the probable consequences of "sharing a joint" is that they exposed a crisis of 'comprehensibility' which both impeded 'rational discourse' and raised questions in the Committee's eyes about the 'truthfulness' or sincerity of the delegation in presenting their remarks. Although the BCCJA representatives tried to clarify their criticisms of the bill adequately enough to be understood by the Committee, the Members either lacked the experience to grasp the relevance of their comments, or they feigned incomprehension in order to deprecate the views of the delegation on the sacrosanct drug plank of the bill. In either case, the exchanges between the Committee and the delegation were punctuated by chastisements of the latter for raising the marijuana example. The end result was, at best, an exaggerated misunderstanding instead of the "communicatively mediated reality" that the hearing event was purportedly aiming to achieve.

c) "Understandability"

If what is 'understandable' depends upon the definition of what is normatively appropriate in a given situation, the House debate at Third Reading[14] epitomized the contention over 'relevant norms' implicit in the Standing Committee hearings on Bill C-36. Amongst the many voices raised in opposition to the bill, one Member debunked its claims to legitimacy on the grounds that it was a "hoax."

> The timing of the introduction of this bill is interesting. Obviously, the government is wishing to be re-elected. If it takes a firm law and order stand, it thinks its chances of being re-elected might be heightened somewhat.[15]

Another Member, doubting the propriety of Bill C-36 as a reflection of the normative mosaic that it was the legislators' duty to uphold, stated that

> we are the legislators. We have to go beyond the initial animal instinct to strike back at someone who strikes at us. We have a duty to the people of Canada to think through what we as legislators are going to do in the legislation that is going to affect Canada for some good number of years to come. Have we done it in the Corrections and Conditional Release Act or have we simply thrown out something to those who are crying for blood? I believe we have only thrown something out. We have not got to the root cause.[16]

The repudiation of government legitimacy in relation to Bill C-36, however, was most emphatically registered by David Barrett, an NDP ideologue and one of the more formidable speakers in the House. Beginning in typically humorous but cutting fashion — "Madame Speaker, I would like to say at the outset that in speaking to this particular bill, I am handicapped by the fact that I actually have some knowledge about this subject" — Barrett identified the 'relevant norms' that had not, it seemed, influenced the shaping of Bill C-36, thus impugning the 'truthfulness' or sincerity of the Government's intentions.

First of all, we have to understand in dealing with this particular problem that we are really defining class.... We currently have a situation where money counts when it comes to criminal justice. The people who end up in jail are generally poor.... We have in this country tens of thousands of young people who we hear about during Question Period who are not even adequately fed, let alone nurtured and sufficiently concerned about to ensure a minimum of criminal activity as a cause of neglect. I am not suggesting that poverty is the only reason for criminal activity, but I am saying that a poverty background breeds the potential for the development of alienation which ultimately leads to criminal activity.... We actually have institutionalized in North America a welfare family system that is known in the United States as an underclass and here in Canada just blandly as the poor.... The consequences of our neglect of the economic and social problems relative to this group of people is that it is a breeding ground for a continuous supply of deviant behaviour by very young people. How does a government morally justify a high level debate about some minor corrections in the Corrections Services when in actual fact it avoids discussing the real problems that exist in real lives in every part of this country?.... This bill is nothing more than window dressing.... In the last 50 years we have never seen such devastation to ordinary folks at the low end of the scale as we have under a laissez-faire, rightwing government that is neo-Conservative and comes in with this kind of clap-trap bill to suggest somehow it is dealing with the problem.... To burden this House with a bill that is fraught with all kinds of political jargon about somehow changing things in our society is to make a mockery of the whole system that we say we are committed to as politicians.... If we really are concerned about the protection of people and property, crime must be prevented by being concerned about the causes of crime and the people who are involved and the victims.[17]

Of course, once this class-war oratory was spent, the Government members voted in unison for the Bill, defeating all Opposition amendments. No Government member contested the logic of Barrett's argument. The 'relevant norms' that he specified were simply regarded as inappropriate within the political parameters of government 'understandability.' Rational discourse, if it prevailed at any stage of the Committee hearings and House debate, certainly ended at the moment of judgement.

IV - Discussion

The examples cited above hardly attest to the impartiality of the State in mediating the 'validity claims' that are brought before it, even in a forum devised specifically for that purpose. There is evidence, instead, that the government's actions are driven by hegemonic requirements — the need, in this case, to uphold the legitimacy of the State while directing and controlling its source of challenge. Given this instrumental motivation, the impact of the hearings — though perhaps tempering the views of Committee Members — is largely inconsequential if the government is already committed to an agenda that does not invite contradictory input.[18]

One may wonder why dedicated, if not officious, senior civil servants in the various Ministries fail to insist on hearings that are more responsive to the plurality of views that come before the Committee. As already noted, the time-frame of Ministers usually calls for quick answers to quickly-conceived problems, so there is little patience for seeking fundamental solutions, especially when Ministers themselves circulate so rapidly within Departments that they have difficulty absorbing the details of the substantive issues in their own area. Moreover, research that contradicts government policy is often ignored, so civil servants learn to put issues in the bottom drawer.[19] At the present time, however, there *are* increasing expressions of concern in official circles about whether more prisons and longer periods of incarceration ought to be viewed as the solution to Canada's increasing crime rate.[20] Even the recent Report of the Standing Committee that presided over the Bill C-36 hearings placed heavy emphasis on "crime prevention" through "social development" in addressing the underlying factors associated with crime and criminality.[21] However, the reliance on "safer communities" and "community policing" as antidotes to crime threatens to restore a punitive focus. Indeed, the language about "crime prevention" tends to fix it at a local, voluntary level, not where its impact might be greatest — the level of massive state intervention. The suspicion is that crime prevention through social development will not be vigorously pursued until the government recognizes its own contribution to the root causes of inequality and crime, an unlikely eventuality.[22]

Of course, the dissatisfaction with the Standing Committee hearing — from the point of view of social and legal reform — can be attributed to the routine imperatives of Parliamentary tradition. Committee Members are appointed by Party Whips, and government members are usually hardliners who seldom deviate from government policy. Whether they allow amendments from the Opposition, or attempt to discredit witnesses' testimony, or deliberately shorten the meeting-time of the Committee or the witness list, depends on how central the bill is to the government's political life. In any case, it almost never happens that a bill is defeated if the government has a majority.

In view of the above limitations, it is not surprising that hearings, as presently planned and conducted, do not contribute very much to the investigative and adjudicative process. As one long-term observer of the Parliamentary scene told me, "They are at best farcical; at worst criminal."

Nevertheless, the problems so far cited for the ineffectuality of hearings as staging areas for genuine reform hardly seem insurmountable; but if the impediments to change are deeply entrenched, then structural interpretations may reveal features of the situation not discernible at the level of participants' individual discourse. For this we turn again to some of the ideas of Jurgen Habermas, and then to the "cultural structuralism" of Pierre Bourdieu.

For Habermas, the quintessential problem of contemporary capitalist society is that it is permeated by "instrumental reason" or means/ends rationality, which saturates modern life with impersonal motives of power. Even government is primarily a conduit for the relentless flows of money and power, neglecting to develop the cultural institutions which would set significant limits on media-steered subsystems. This failure or inability to reconcile the competing claims of lifeworld and system-processes precipitates in legitimation and motivational crises, especially when the "system" fails to deliver the goods. The solution, according to Habermas, lies in a resurrection of the "public sphere," bringing a restructuring of meaning and commitment in social life and a restoration of the proper balance between system and lifeworld processes. A revitalized public sphere, then, would reduce 'power-distorted communication' and foster 'rational discourse,' enabling the creation of a more just, open and free society.

According to their professed function, we might anticipate that 'hearings' would utilize lifeworld processes in order to recreate the mutual understandings capable of integrating the social order. This was not exemplified, however, by the hearings on Bill C-36, which seemed to observe a means/ends rationality framed by electoralist opportunism and fiscal considerations. Indeed, this typified the paradoxical function of hearings in *subverting* the very spheres of symbolic reproduction (or lifeworld) they might be expected to uphold, and promoting the assimilation of those spheres into the organized domains of economic and bureaucratic action.[23] Rational discourse, by Habermasian standards, does not prevail in these circumstances, as deliberations are geared to contriving means to attain predetermined ends.

If hearings, then, present as a cultural form that disappoints its own rhetoric, its persistence is puzzling but may be accounted for in the structuralist analysis of Pierre Bourdieu, which is predicated on the assumption of only *one* center of legitimate cultural values within a society.[24]

Bourdieu rejects the conventional understanding of class struggle (whether Marxist or Gramscian) as initiated by the dominated classes to overcome their status of being dominated. Instead, he conceives of class struggle as a symbolic struggle for cultural distinction that takes place not between dominant and dominated classes, "but among dominant and dominated fractions within the ruling class itself" (Joppke, 1986: 67). Socially dominated classes, according to Bourdieu, lack the cultural capital to effectively challenge the legitimate center of cultural values. Rather, they engage in "defensive adaptations" to the objective conditions of their existence (i.e. through expressing their "class habitus" or durably acquired schemes of perception) which induce them to "collude in their own domination" (Jenkins, 1992: 98).[25] Only the dominated fraction *within* the privileged ruling class can challenge the legitimacy of a prevailing 'symbolic classification pattern,' a challenge muted by their interest in improving their own position within the existing stratification order. And only under rare conditions of socioeconomic crisis, when a wide gulf exists between their habitus and objective situation, might the dominated fraction of the elite class work towards a "radical inversion of the table of values." As for the proletariat, they are "...invisibly forced to be silent or to resort to the clumsy and fragmented use of a language borrowed from political and cultural elites" (Joppke, 1986: 68). Indeed, "...by the mere fact of their taking part in this kind of struggle, the existing order and its rules are legitimized and affirmed. In this regard, the competitive struggle is essentially an integrative and reproductive struggle" (*Ibid.*: 75).

In what sense, then, is the hearing occasion elucidated by Bourdieu's analysis? To begin, we might presume that the very regularity of the hearing occasion certifies it as a tested feature of domination. As Bourdieu writes,

> Once a system of mechanisms has been constituted capable of objectively ensuring the reproduction of the established order by its own motion...the dominant class have only to exercise their domination. (Bourdieu, 1977: 190)

In this sense, the hearing is a particularly apt embodiment of the exercise of domination since dissent is aired without any likelihood that it will burst the bourgeois hegemony. Speakers representing the dominated classes lack the cultural capital to insist that legislators redress their grievances, and participants from the dominated fraction within the privileged ruling class, who *are* capable of challenging the legitimacy of the prevailing symbolic classification pattern (as professionals, educators, managers, etc.) entrust their criticism to the legislative will of the Committee, thereby preserving their niche in the one legitimate center of capitalist society while awaiting their own opportunity to dominate, under the vindicating illusion that they will do otherwise.[26]

We can, of course, reject Bourdieu's notion of the reproduction-guaranteeing inertia of habitus, on the grounds that it offers no account of the real oppositional forces in society and reduces social change to a neo-Paretoian version of the "circulation of elites" (Joppke, 1986: 69). But if we wish to understand the continual reproduction of power relations, even in the midst of apparent legitimation crises, then we cannot disregard the implications of Bourdieu's analysis for the understanding of social democracy (in all its legislative trappings) as a profoundly *hegemonic* project.

V - Conclusions

Hypothetically, hearings can be a vehicle for social change. In extending the representation of interests, they provide a counterbalance to government proponents and heighten the profile of state accountability. In practice, however, hearings are largely exercises in 'repressive tolerance,' enabling grievors to let off steam while providing the state with added legitimation. Interestingly, most witnesses who agree to perform at these occasions *are* aware of the misleading nature of the proceedings, yet participate, so they claim, in order to get their views on record for political reasons. Indeed, it is no secret that hearings typically operate under government control, and modifications of Parliamentary process to loosen Party discipline have long been under consideration, with some changes said to be imminent, such as enlarging the role of backbenchers so that they can better represent their constituencies and regions, freeing Members from the expectation that they need vote with the government on every issue, limiting the sorts of motions on which the government can be said to have lost the confidence of the House, and changing the way private Member's bills are dealt with in order to improve their chances of passing. While these changes may occur, and may have some bearing on the conduct and outcome of hearings, it is doubtful that they will significantly alter the key function of hearings as a 'non-coercive' device for preventing the mobilization of dissent. Moreover, if changes to modes of domination are left to those who do the dominating, then *plus ça change, plus c'est la meme chose.* Such changes would merely reaffirm the relative distance between the social classes and maintain the inertia of habitus.

Short of a revolutionary situation that would destabilize class relations, social change depends upon the actions of those with the material and symbolic means to dissect and reject existing definitions of reality. Political struggle, in this context, requires the eschewal of dominant cultural forms that ensure the reproduction of class habitus. 'Hearings' are but one example of the false fronts on which that struggle is idly waged.

Endnotes

1. For a detailed discussion of government objectives in the Bill C-36 legislation (eventually passed), see *Directions for Reform: A Framework for Sentencing, Corrections, and Conditional Release,* 1990. For an evaluative review of the major proposals in Bill C-36 and the inquiries leading up to the new legislation, see Ratner (1992).

2. The discussion of Habermas's work draws mainly on sections of Jürgen Habermas, *The Theory of Communicative Action,* vols. 1 and 2, (Boston: Beacon Press, 1983 and 1987).

3. For Habermas, "Lifeworld" is a culturally transmitted and linguistically organized stock of interpretive patterns with respect to culture, society, and personality.

4. The "ideal speech situation" is one in which actors possess all of the relevant background knowledge and linguistic skills to communicate without distortion.

5. In utilizing some of Habermas's concepts for the purpose of clarifying the actualities of such hearings, I do so without attending to the not insubstantial criticisms that have been made against the ontological foundations of Habermas's work; that, for example, his theorizing is rooted in patriarchal assumptions, signified by his neglect of 'emotions' (Meisenhelder, 1989: 126-132), and that it is irremediably naive or romantic because it establishes criteria for communicative competency that, in *any* era, would be unattainable (Turner, 1991: 280-281). These criticisms, though arguable, do not detract from the application herein.

6. The legislative process involves a number of steps, starting with a Ministerial initiative (usually recommended by Department senior staff), Cabinet and Treasury approval (the latter if expenditures are required by the new legislation), further approval by the Priorities and Planning Committee of the Government, First Reading in the House, Second Reading, referral to the appropriate House Standing Committee for clause-by-clause review and possible amendments, Report Stage Debate, House Third Reading, and, if the legislation passes,

repeat of most of the same process in the Senate. The House Standing Committee Hearing, the focus of this paper, is where we would expect the most thoroughgoing examination of any bill to occur, assuming that the Government wishes to submit its initiatives to genuine scrutiny.

I am thankful to Gary Levy, Editor of the *Canadian Parliamentary Review*, David Miller, former Executive Assistant to John Turner, and Penny Reedie of the Security and Intelligence Secretariat, Privy Council Office, for their lucid explanations of the vagaries of the legislative process.

7. This proportion of political party members is generally the same for all Standing Committees (5,2, and 1), with a quorum of five. The government party would always have a committee majority in a majority Parliament.

8. Canada, Parliament, House of Commons, Standing Committee on Justice and the Solicitor General, Minutes of Proceedings and Evidence, no. 32 (9 March 1992) (Ottawa: Queen's Printer, 1992), p. 34.

9. *Ibid.*, no. 33 (9 March 1992), p. 7.

10. *Ibid.*, pp. 23, 24.

11. *Ibid.*, p. 39.

12. *Ibid.*, p. 43.

13. *Ibid.*, pp. 47, 48.

14. Recorded in House of Commons, Debates, vol. 132, nos. 137, 139, 140 and 142 (7, 11, 12 and 14 March 1992) (Ottawa: Queen's Printer, 1992).

15. *Ibid.*, no. 140, p. 10596.

16. *Ibid.*, p. 10608

17. *Ibid.*, no. 142, excerpts from pp. 10704-10707.

18. Proof of this imperiousness was the fact that Bill C-36 was reported to the House by the Committee for Third Reading despite the fact that many organizations (as well as the Opposition in the House) urged that the government hold the bill in abeyance until its provisions could be meshed with those sections of the bill that would refer to sentencing (eventually, Bill C-90). This was not observed, although the Committee Chair said he would do so during the Vancouver hearings. Apparently, the government did not want to wait any longer to provide a political sop to the electorate.

19. I am indebted to Barry Leighton, a senior research officer with the Department of Solicitor General, for sharing his observations with me on these points.

20. The recent meetings of the prestigious Society for the Reform of the Criminal Law questioned the use of the criminal code to deal with what were seen to be essentially social problems (*Vancouver Sun*, 3 July 1993). Even Bob Horner, the Standing Committee Chair, had second thoughts about a 'hang 'em high' attitude, stressing the social roots of criminality only a short while after the passage of Bill C-36 (*Globe and Mail*, 10 March 1993). The sincerity of his

reflections must be matched against the endorsement of sitting MPs (for the next federal election) by the Canadian Police Association. Horner was one of those endorsed, presumably because of his tough law-and-order stance (*Vancouver Sun*, 19 July 1993).

21.   As one example of the Report's attention to root causes, it is stated that, "These accounts of the conditions that contribute to crime and criminality make clear that there is no single root cause of crime. Rather, it is the outcome of the interaction of a constellation of factors that include: poverty, physical and sexual abuse, illiteracy, low self-esteem, inadequate housing, school failure, unemployment, inequality and dysfunctional families...crime cannot be prevented solely by the criminal law and criminal justice services. It is a social problem that requires all sectors of society to work together for safer communities" *Crime Prevention in Canada: Toward a National Strategy*, Twelfth Report of the Standing Committee on Justice and the Solicitor General, Dr. Bob Horner, M.P., Chairman, House of Commons, *Debates*, Issue No. 87, February, 1993: 11-12.

22.   In commenting on the government's crime prevention rhetoric, a senior staff member at the Department of Solicitor General, who had best remain nameless, described himself as "a participant-observer in a gigantic farce."

23.   For an excellent synopsis and application of Habermas's concepts, see Malhotra (1987).

24.   "...*all* symbolic practices, utterances and beliefs have to be measured by this relative 'distance' to this imaginary focal point." (Joppke, 1986: 67)

25.   'Habitus' is the mediating process between class position and individual behavior; the structured system of dispositions which defines the propensity to act in specific, regular sorts of ways; it constitutes a 'practical sense' or 'feel' for the social game; the basic unit of habitus is the individual actor as a member of a social group or class.

26.   On this point, I recall the strident criticism of government operations and secrecy by Derek Blackburn, NDP member of the Standing Committee. Following the hearing on Bill C-36 and the issuance of the Committee Report, Blackburn decided to quit politics for an $85,000 per year government job, which would also allow him to collect his MP's pension simultaneously ("double-dip"). It is these rewards and enticements that blunt the criticism of potential trouble-makers, ultimately converting them into agents of hegemony and demonstrating the constraining effects of habitus.

References

Bourdieu, Pierre. 1977. *Outline of a Theory of Practice,* Cambridge: Cambridge University Press.

Habermas, Jurgen. 1983, 1987. *The Theory of Communicative Action,* Vols. 1 and 2. Boston: Beacon Press.

Jenkins, Richard. 1992. *Pierre Bourdieu.* London: Routledge.

Joppke, Christian. 1986. "The Cultural Dimensions of Class Ferment and Class Struggle: On the Social Theory of Pierre Bourdieu," *Berkeley Journal of Sociology.* Vol. XXXI, pp. 53-78.

Malhotra, Valeria Ann. 1987. "Habermas' Sociological Theory as a Basis for Clinical Practice with Small Groups," *Clinical Sociology Review.* Vol. 5, pp. 181-192.

Meisenhelder, Thomas. 1989. "Habermas and Feminism: The Future of Critical Theory" in Wallace, Ruth A. (ed.), *Feminism and Sociological Theory.* Newbury Park: Sage Publications.

Ratner, R.S. 1992. "Bilateral Legitimation: The Parole Pendulum" in Livy Visano and Kevin McCormick (eds.), *Canadian Penology: Advanced Perspectives and Research.* Toronto: Canadian Scholars' Press Inc.

Turner, Jonathan H. 1991. *The Structure of Sociological Theory*, 5th ed. Belmont: Wadsworth Publishing Company.

Community Programming for Young Offenders in Ontario: From Neglect to Rhetoric

Thomas O'Reilly-Fleming and Barry Clark

Introduction

The past decade has seen a fundamental shift in the legal system's treatment of young offenders in Canada. The passage of the *Young Offender's Act* (*YOA*) ushered in a new era of conditioning of adolescents more in keeping with Foucault's notion of the production of the docile body than with the assessment and response to offender needs. Indeed, it may truly be argued that the 1980s reflected a renaissance of the 'exceptional state' (Ratner and McMullan, 1985) in Canada, and a return to a 'just desserts' model of juvenile processing by our criminal justice system. In this movement back into the carceral concerns of earlier eras, the *YOA* signalled its retrogressive character, and its focus on correctionalism rather than on rehabilitation. The 1960s had been a decade in which we emerged from postwar conservative dialogues and their attendant philosophies which ensured that offenders, both adult and juvenile, were subject to at times arbitrary, and often harsh and lengthy sentencing (Caputo, 1987; 1991).

The passage of the *YOA*, although fraught with some measure of debate upon the direction of youth legislation, produced a new consensus. This consensus concurred with the failure of the rehabilitative ideal and emphasized the unholy triumvirate of crime, punishment and responsibility. References drawn from the House of Commons debate which accompanied the passage of the Act into law clearly demonstrate a dramatic shift in the intent of the law. However, the *YOA* is also concerned with some very fundamental precepts and philosophies, guidelines which were meant to directly effect both the spirit in which the sections of the Act were to be applied and the approach that would be taken with those young people who found themselves in conflict with the law.

In this chapter we concern ourselves with the fundamental gap that has arisen in Ontario between the spirit of the *YOA* and its translation — or more accurately lack thereof — into community programming. We wish to argue that Ontario has cemented itself within the foundations of a correctionalist mode of dealing with young offenders which has an almost exclusive focus upon incarceration. First, we present an analysis of the spirit of the *YOA*. Secondly, we trace the carceral preoccupations of the current system. Finally, we describe several community corrections programs with the potential to provide a positive response to juvenile offenders. We argue that creative responses must take into consideration wider socio-economic forces that are at the heart of deviance production. Further, it is suggested that current efforts fail to address needs of class and race whether framed within debates concerning the failures of economic marginalization, at the level of control agency interaction or within the correctionalist "net" (Cohen, 1985).

The Spirit of the YOA

Canada lived under the spectre of the *Juvenile Delinquents Act (JDA)* (1908) for some eight decades. While the Act had some redeeming features, it had by the mid-1970s outlived its utility in the Canadian context. The reasons for the decline in popularity of this Act are important to understand in order that the remainder of the argument presented in this chapter is contextualized. The *JDA* emphasized the spirit of *parens patriae*, that is, the state as parent. The judge was the centre of a patriarchal system that saw him as substituting for the failed parents of delinquent youth. While parents were reduced to a state of incompetency (and hence blameworthiness) under this legislation, nonetheless it was intended that the parents play a significant role in rehabilitating the young offender to full civil status. The temporary failings of the parental dyad permitted the intrusion of the state in order to address the 'needs' of the young person, for delinquents were seen as persons who suffered from some form or combination of social, educational and certainly moral deficits. Under the *JDA*, children's behavior within the moral sphere — so-called status offenses — and behaviors including truancy, and running away from home were punishable by imprisonment in "reform" schools and a variety of other punishments. The *JDA* was riddled with hypocrisy and inequity as juveniles were sentenced to longer periods for offenses than adults convicted of the same crime, were sentenced for behaviors that were not offenses when committed by an adult, and were subject to provisions that permitted them to be returned to reform schools even if they had committed no further offense. Sentencing was also of an indeterminate nature so that young people could spend a considerable amount of time in custody if their behavior was adjudged to be unchanged since the adjudication of their offense(s).

While the underlying philosophy of the *JDA* stressed a needs-based approach, there were fundamental differences between the philosophy in word and the actual delivery on its promises. Reform schools were loosely concerned with rehabilitation after the 1960s but certainly throughout their history they stressed individual punishment and reform. While the atmosphere of juvenile courts evidenced all the trappings and circumstances associated with justice and a concern for the welfare of the young person, there was in reality a pervasive form of patriarchy operating under the surface. Juvenile court judges were political appointees who typically had enjoyed long service in the social work field and were trained social workers. They viewed themselves as literal fathers (and later mothers) to young people but in a way that is consistent with the child that is being reprimanded for their behavior, even when the hollowness of the law being applied was obvious to the object of its sting, the delinquent. Parents and educators were permitted to use the court as a form of punitive backup for their own disciplinary ideas, or were content to go along with the court's approach if their own had failed. The courts dealt primarily with disaffected working class youth until the 1970s — individuals who were unlikely to seek legal counsel. In a very real sense the operation of the court relied upon an approach characterized by informality. Judges spokes directly and often to the accused, parents or guardians and social workers concerning the youth's behaviors and attitudes. While on one hand this might seem to have signalled a benevolent approach, this turned to vapour at the time of sentencing. Reform school sentences were real, hard and ofttimes extended, a fact that few juveniles were cognizant of. Even by the 1980s, many juveniles were reported to be experiencing significant difficulties interpreting what was going on in the courtroom (Smith-Gadacz, 1983).

By the 1970s there were significant reforms that were being contemplated at the juvenile court level prompted by several important social trends within society, in our opinion. First, the movement within social work and social sciences towards the "rehabilitative ideal" was having a considerable impact upon the justice system's view of its task. There was marked and prolonged debate concerning the use of the carceral system for purely punitive ends. Secondly, broader social movements towards individual freedom and rights with the civil rights sector in the United States, within the womens' rights movement, and emanating from the "hippie" revolution had a spillover effect within the juvenile justice system. One of the telling lessons of the late 1960s was the need for legal representation to ensure an individual's rights and the conduct of courts with regard to due process issues. In Canada, several groups dedicated to the procurement of justice for children sprang up in the early 1970s, and academics as well as other researchers began to pay increasing attention to the treatment of juveniles under the *JDA*.

Few analysts at the time or since have understood the centrality of this shift. Judge Lorne Stewart of the Toronto Juvenile Court was cognizant of the changing nature of legal relations within the juvenile court system and recognized the need for new appointees to have the requisite legal training to cope with an increasing number of young offenders who came to court with legal counsel in tow. Counsel were intent on technical argument which proved increasingly to be beyond the capabilities of the existing judiciary.

Halfway Houses

One example worth exploring which characterizes most clearly the growing mandate for new approaches to dealing with delinquents during this period is the halfway house movement. It provides somewhat of a link between the old approach of the *JDA* and foreshadows the early development of the *YOA* in Canada. The 1970s also heralded the beginning of a substantive shift in the focus of our approach to the 'correcting' of young people. The *halfway house*, for young people who were suspended, as it were, between the carceral and the community began to emerge as an alternative to imprisonment. Opportunity House, in Toronto, Ontario was the first co-educational, co-operative group home for young persons in trouble with the law to open in Ontario. It was at the cutting edge of innovation with regard to the role the community could play in rehabilitating young people with an emphasis on group techniques and individual counselling within a structured environment as basic elements in addressing the needs of young people. Gradually, and in small increments, alternatives to incarceration developed utilizing the group home approach. The young offender benefitted in several ways from this arrangement. First, it permitted them the opportunity to avoid entrenchment in the carceral network, which, as research readily confirms, exerts a centripetal force on its unwilling inhabitants, drawing them back, again and again. If recidivism were not a strong enough argument for the need for innovatory measures then certainly the costs — both financial, and more centrally, human — of the carceral enterprise added additional support to calls for the expansion of community programs. Secondly, the halfway house allowed the young person to remain either in school, training or in employment, a less disruptive alternative when compared to imprisonment. Thirdly, this approach allowed young people a form of asylum from the familial and other contingencies that brought them into conflict with society. Finally, group homes offered a forum for young people to exchange ideas and information with a peer group that had similar experiences, and so informed their development, an early form of reality therapy.

From Rehabilitating Juveniles to Rehabilitating the Law

The birth of the *YOA* has been extensively documented in the Canadian context by Caputo (1987), O'Reilly-Fleming and Clark (1992) and a variety of other researchers, and so we will not belabor the reader with a descriptive analysis of its inception and development. However, there are some significant signposts that it is worth highlighting in order to buttress our coming analysis of recent community innovations in treatment. The *YOA* has come under increasing criticism from a number of groups within society who criticize the act on a number of substantive issues. On the anniversary of the first decade of the Act's operation police are mounting a continuing demand for the dropping of the age at which young offenders can be charged back to seven years of age. One activist group in Ontario wants schools to be able to accumulate records of the offenses that juveniles have committed during the seven to 11 age range. This, like many other criticisms of the *YOA*, is founded on the belief that violent crime is rising, and that young children are monsters in the making whom the criminal justice system must have access to in order to effect early, prompt and significant change in the child's relation to society. However, ideas of eight-year-old rapists, murderers and thugs are far-fetched and not supported by the empirical data. National data on youthful offending collected by The Centre for Criminal Justice Statistics clearly informs us that those in the seven to 12 age range are primarily involved in mischief offenses. In fact, serious crimes by this group, which is extremely small to begin with, rank at the .05 level. It is arguable that these "offenses" would even appear in police records were it not for the widening of the net of control to include children whose acts might be dealt with effectively in a much more informal manner *sans* police involvement. Other lobbyists decry the *YOA* as 'soft' on young offenders when there is clear evidence that young offenders are receiving far more sentences under the *YOA* than would have previously been the case under the *JDA*.

While the *YOA* represents a compromise document that is alternatively referred to as a Liberal or Conservative document depending upon whether praise is being given or withheld, it represented a significant shift in our philosophy towards the treatment of young people. The underlying principles of the *YOA* are laudable, and in context they appear to provide a deepening of our commitment to the needs of the offender. They include: acknowlegement of the state of dependency and level of development and maturity; recognition of the special needs which require guidance and assistance; special rights and freedoms; policies which provide the least possible intrusion into freedom of the young person; options which stress non-intervention or other than judicial proceedings; and attempts to have parents continue in a supervisory role wherever possible. Overall, in context, these foregoing principles are to be liberally construed. While these underlying principles of the *YOA* and concomitantly the *Child and Family Services Act* are to provide, in tandem, a recognition of the special state of childhood, there are clear signs that we analyze in this chapter that this is not occurring. Whereas the best interests, protection and well-being of children is to be the paramount feature of the approach under this latter legislation, there is an inconsistency between these broader principles and the stated intent of the Act. The introduction of the Act in Parliament and the debate surrounding it, was, with few exceptions, one that focused upon the twin elements of punishment *and responsibility.* This marks a radical departure from the spirit but not the practice of the *JDA* in many respects towards a view that emphasizes the *protection* of the community rather than the *salvageability* of the delinquent.

We have argued elsewhere that such an approach is inextricably wedded with the concept of incarceration as the only viable means to enforce the law, achieve deterrence and teach the young offender about his responsibilities as an errant citizen. This is consistent with Foucaultian notions of the need to place the prison in a central position in the state control apparatus. It is, despite the rhetoric, we argue, largely a return to older notions contained within the crime-responsibility-punishment framework previously abandoned in the 1980s as an overly punitive, nonresponsive and criminogenic response to the criminality of young people. Morevoer, research from a variety of quarters has amply demonstrated that not only are charges rising under the *YOA* (Juristat, 1991) but that many more youths are being sentenced to longer periods of incarceration or probation than was previously the case under the *JDA* (Leschied and Jaffe, 1989). A balanced review conducted by Tony Doob (1992: 82) of The Centre of Criminology in Toronto found that "...there has been an increase in the proportion of those cases getting to court that are receiving custodial sentences." These are likely to be "shorter" sentences of three months or under, a form of the "short, sharp, shock" — that is, sentences which are meant to impress upon the juvenile the seriousness of their crimes but remove the offender before the effects of longer term incarceration begin to become manifest.

Locked into Custody: Ontario, Incarceration and Opting Out

The YOA is fraught with considerable difficulties in terms of treating young offenders who are suffering from emotional, behavioral or psychological deficits. Under Section 22(1) a young person who is in need of psychiatric intervention retains significant powers in terms of approving treatment regimes. Under this section the approval of the parents, the treatment facility to accept and the young offender are all required before treatment can be offered. Young offenders retain the right to "opt out" of treatment at any time after the commencement of efforts.

Significantly, Ontario decided to 'opt out' of certain key provisions of the YOA during almost the entire first decade of its existence. First, Ontario decided to refrain from providing researchers and policymakers with statistics on the criminal activities of young people, thus making any assessment of the operation of the YOA suspect, given that the most populous province was not co-operating with data collection efforts. Secondly, Ontario decided to continue to make its investment in custody beds rather than in the development of alternative measures programs in concert with other provinces. This has meant effectively that the spirit of the principles was thwarted by the ability of Ontario to refuse to participate in providing meaningful alternatives. Therefore, the concepts of the interests, protection and well-being of the child have been buried, literally, in the brick walls of Ontario's ever-expanding and increasingly punitive juvenile carceral network.

This has meant that Ontario has, and continues to make, a massive investment in the carceral system spending some 90 percent of available funds on institutions and virtually no money on community alternatives. Historically, the building of institutions has been folllowed by the influx of deviantized clientele. Jones (1960) in her analysis of England's most ambitious period of institutional construction during the eighteenth century found that increasingly marginal forms of behavior became the subject of legislation, medicalization and deviantization, and hence, relegation to the madhouses and prisons of the time. The boundaries of what constituted deviant behavior constantly expanded as more institutional places were created. Ontario's decision to proceed with a system of warehousing its juvenile offenders is no more than a short-term solution to a growing societal problem with young people who find themselves disaffiliated from broader social forces. Youth unemployment is, in and of itself, a major factor in the production of youthful criminality in Canada. Elements of racial isolation and mistreatment have had a part to play in the production of much recent highly publicized deviance by youth gangs (Lewis, 1992; O'Reilly-Fleming, 1993). The majority of youthful offending in Canada revolves around simple crimes of property. Caputo (1987) as well as a number of leading critical scholars in the field of delinquency research in Canada have informed us that some of the central contributory themes of delinquency production have been either totally ignored by researchers, or at the very least, have received scant sociological attention. A brief list would include:

(1)　Societal practices and ideas concerning the lengthening of the ages to which adolescents remain dependent in economic terms. This argument is most cogently developed by West (1984) who views this form of economic partialism, in which the youth is restrained from well-paid employment (by virtue of age and education) while being encouraged in part-time marginal employment as a prime cause of youth disaffection.

(2)　Treatment of youths as dependents by the educational system which extolls the benefits of commitment to exclusion from the labor force over a protracted period in exchange for meaningful employment later in life. Unfortunately, trends within the Canadian economy have produced a large underclass of educated and skilled individuals who are without employment. Arguments concerning the rewards of awaiting those who conform and meet educational expectations are often viewed as hollow, serving rather transparently the ends of social control they were intended to (O'Reilly-Fleming, 1993).

(3)　A more contemporary theme arises from the isolation and exclusion of racial minorities drawn from recent immigrant classes from the economic and educational mainstream. While one cannot argue that this is a universal trend, there is some evidence that gang membership, for example, is promoted in immigrant youth because of difficulties in "fitting" with the dominant cultural values. Feeling an outsider in the educational setting, the street gang assumes the role of the family for young people, as much as street associates for adolescent homeless fulfill the roles intended for the familial group.

(4)　Youth express feelings of disenchantment with the dominant forces in contemporary culture which has "centred" upon the 1960s and its "hippie" revolution as reminiscent cultural signposts.

Popular music, aside from rap which mirrors contemporary black experience in North America, derives from 1960s music. It imitates the form popular at that time or presents "60s survivors" who still form one of the dominant forces on radio, tracking the baby boom listening crowd. The "collective search for identity" is much more fragmented than those distinct forms available in the 1940s, 1950s or 1960s popular culture.

These underlying trends find reflection in Ontario's increasing use of custodial force under the *YOA*. At the root, the use of brute force has assumed a variety of clear forms which are worthy of consideration:

(1) There is evidence of systemic racism within the entire Ontario criminal justice system according to Attorney General Howard Hampton who ordered a three and one half million dollar review of racism in the Ontario context in October 1992. This review applies equally, if not more centrally, to the *YOA* and its ancillary systems and constitutes one form of response to the Lewis Report.

(2) The use of custodial sentences at the expense of alternative measures signals a deep commitment to ignoring needs and concentrating on punishment.

(3) Despite movements in other legal jurisdictions in the Western world towards decreasing formalized contacts with juveniles, Ontario has behaved like a disciplining parent who is doomed to fail since constant intervention constantly produces ever greater numbers of offenders. According to Cohen (1985), the net continues to grow and the fish palatable at the correctional table continue to diminish in size, but not in importance to the correctional enterprise. This is particularly true in a system that has shown signs of moving towards privatization of control (Ericson *et al.*, 1987). In Britain, police have made use of cautioning procedures whereby juveniles, rather than being put through formal arrest, charge, jail and court procedures at considerable expense, spend time at the police station, are interviewed, cautioned and counselled regarding their offense. While such a proce-

dure is obviously inappropriate in the case of serious crimes one can see some great degree of potential for removing juveniles from the formal control system in Ontario, through the institution of this form of program. Combined with a victim-offender restitution, attendance centre or COPS program (discussed later in this chapter) it would divert thousands of young people out of a custodial system.

(4)   Despite reforms and well-intentioned critiques, a large percentage of young offenders are the victims of carceral inertia, that is removal into the YOA archipelago, and out of society largely as a result of economic disadvantage or bordeom-relieving behaviors. Over a decade ago Turk (1985) pointed out that such crude methods do not have sustainable results, and furthermore, the results achieved are typically not those desired.

(5)   Punishment-based systems are not productive and are prone to corruption. No one in Canada can argue in the wake of Mt. Cashel and a number of other similar incidents that institutions are fit or trustworthy repositories for young people who typically evidence grave needs on a variety of fronts.

(6)   The split-jurisdiction model which divides offenders on age criterion between Community and Social Services and Corrections has created inequities in sentencing, and lack of ministerial co-operation in the best interests of the child.

What Is to Be Done? Three Programs

Native Programs

One of the values of the Community Options Program (COPS) is its versatility. Somewhat chimerical in nature, its curriculum can be adapted to a variety of dissimilar situations. A promising example of this flexibility is found in its application to the aboriginal context, even isolated native reserves. The North Caribou Lake Band model, for instance, is a regional/cultural modification of the Southwestern Ontario program. Both the content and the process of the program have been adapted to meet the needs and requirements of the isolated northern Ontario reserve. During the past decade many sequestered reserves have undergone a McLuhanesque upheaval of their traditional religiocultural patterns. Satellite television, while introducing a new and immediate access to world events, has widened the knowledge gap between native elders and youth and between traditional values and modern norms. High rates of alcoholism, family violence and school dropout rates, as high as 80 percent in some reserves, dramatize the particular plight of the aboriginal condition. An external justice system, from policing to a visitant circuit court, reflect the hegemony of the dominant white power structure. Lengthy delays between the commission of offenses and their disposition through court — many of the dispositions are virtually unenforceable — dilute the authority of the formal justice system. While the Community Options Program (explored in detail in the next section) is not a panacea for the hybrid and entrenched problems experienced on the remote native reserves, it is, in its adapted form, a highly responsive agent of prevention at a variety of levels. For instance, the standard basic curriculum developed for native reserves reflects a cultural and geopolitical sensitivity to a range of endemic problems. Fundamental units of the curriculum, incorporating elements of community and personal service, include sports nights, remedial homework programs, community work options, a junior firefighters program, firearm safety (hunting/ trapping), fire safety, skidoo/ outboard motor maintenance and repair, first aid and household safety programs, a native artist program, quilt making/beadwork/soapstone carving/ woodwork, music, native spirituality, babysitting/parenting programs, drug/ alcohol health promotion, understanding the media, anger management/family violence, teen dances, teen radio programs and more.

Part of the broader socio-political process is to establish and to work closely with a native policing and native counselling component. Youth Justice Committees, comprised of band elders, provide leadership and ensure aboriginal ownership and empowerment over the system. Ideally, the Community Options Program, in the native setting, manifests the principles of the *Young Offender's Act* through a continuum of applications from the pre-charge voluntary (under 12s), the Alternative Measure Intake (A.M.I.), Intensive Probation Supervision and, importantly, a post-custodial function. These measures are particularly helpful when used as either an alternative or a follow-up to custody, since incarceration, for natives, primarily entails removal from the community.

The Community Options Program

The Community Options Program (COPS) was conceived in 1979 under the initiative and supervision of the late Herbert O. Kennedy, senior Probation Officer in the Waterloo Area Office of the Ministry of Community and Social

Services, Waterloo, Ontario.

The model was developed and expanded to Waterloo, Wellington and Grey/Bruce Counties as a contracted service to the John Howard Society of Waterloo Region. This COPS model will be briefly examined, in the context of the preceding issues, since this program addresses many of the problems that plague the implementation of the *Young Offender's Act* in Ontario.

The COPS model contains the following positive features:

(1)   it faithfully reflects the principles of both the *YOA* and the *CFSA*;

(2)   the program curriculum is adaptable to the age and developmental needs of both children and young persons;

(3)   program units can be tailored for cultural, ethnic and gender differences;

(4)   the program can respond to a continuum of judicial/correctional need (i.e., pre-charge voluntary, postcharge/pre-dispositional, post-dispositional and postcustodial models;

(5)   COPS is cost-efficient; and

(6)   overall the program has been endorsed by both private and governmental sectors and reduces the problem of overcrowding in young offender institutions.

COPS comprises three service areas:

(1)   community service orders wherein the offender works within a community agency to fulfil his sentence;

(2)   attendance centre programs which permit the offender to benefit from counselling; and

(3)   primary prevention services. This final component includes an array of offense-specific programs including anti-shoplifting, family violence and drug/alcohol abuse prevention units. The curriculum kit includes workbooks, videos, overheads and board games. The units are presented in elementary schools, generally at a grade six level, and as part of the province's Values, Influences and Peers (V.I.P.) initiative.

As well as being directed at a general school population, a more intensive curriculum can be introduced into high-risk schools and linkages created between complementary programs such as Student Watch (school anti-vandal-ism) or S.A.D.D. (Students Against Drunk Driving). The units are presented in elementary schools, generally at a grade six level and as part of Ontario's Values, Influences and Peers (V.I.P.) program as well as being directed at a general school population. As well, individual referrals can be made from the commu-nity, schools and law enforcement, usually at a voluntary level, with a special focus on the "under 12s."

The Attendance Centre model is the intensive and adaptable component of the COPS program repertoire. Its application, with appropriate curricular modifications, can be pre-charge/voluntary (usually for the under 12s), post-charge/pre-dispositional (perhaps as a formal Alternative Measures), post-dispositional (usually as a term of regular probation) and, finally, postcustodial (as an early release or standard reintegrative followup to custody). The Attendance Centre is essentially a highly structured and intensively supervised correctional intervention. Children and young persons attend an evening program, voluntarily or mandated, up to five evenings a week. Normally, they are transported to and from the program by staff, between the hours of six p.m. to nine p.m. On occasion, post-program curfews and weekend activities are added to extend the program's supervisory strictures. On the average, 10 participants attend each evening; however, the average nightly program count may be double that number since many participants will be in a part-time phase-in or phase-out stage or in a Community Placement Program. This placement program arranges for community activities (hockey, swimming, guides, leader-in-training etc.) as a monitored part of the regular Attendance Centre program. It is also used to extend the realm and duration of regular program involvement after participants graduate from the Attendance Centre. The average program Involvement is six months, although, based on individual progress, this duration can be shorter or longer. An Attendance Centre is usually staffed by two full-time and one part-time professional, normally a combination of one accredited teacher, and graduates in the social sciences and recreation. Referrals come from parents, youth agencies, the schools, the police and the courts, depending upon the formal or informal nature of the referral process. An intensive assessment period always commences program involvement and will dictate the length, intensity and nature of the individual's program curriculum. The Attendance Centre, which operates nightly out of schools and community recreational facilities and parks, offers a broad and attractive program menu for participants. While some curricular units are mandatory, there is a heavy emphasis on a voluntary, participant-selected options. A typical Attendance Centre program curriculum, developed and delivered by the professional staff, would include: offense-specific modules (anti-shoplifting, antivandalism), family violence, anger management, self-esteem, respect for authority, racial/cultural appreciation, nutrition, hygiene, study skills/homework completion and employment/job placement. A full range of less formal units are also standard program fare: recreation, crafts, computers, woodworking, welding, as well as certification programs in St. John's Ambulance, C.P.R., swimming, babysitting, bicycle safety as well as tours of community sites. Many students who are placed on daily homework journals, receive individualized tutorial assistance at the Attendance Centre where time is set aside for homework completion and exam preparation. The homework is read and signed daily by the participant's regular teacher. Participants may also be placed on individualized behavior contracts in the home and/or at school which are monitored by the Attendance Centre staff daily or weekly. It is highly recommended that, in some instances, a family therapist, attached to the COPS program, be assigned to high-risk, multi-problem youth and their families, and should include an assessment referral component. As earlier mentioned, all participants are encouraged to join community activities and organizations such as formal swim classes, volleyball, baseball, etc. Attendance at these placements is monitored and is arranged through the Attendance Centre with funding often coming from local service clubs. The placement program continues and often intensifies once a participant graduates from the Attendance Centre. It is not surprising that, with such varied curricula scaled to the needs and interests of the participants, regular attendance is in the 90th percentile. Admittedly, attendance and success is greater for the younger participants, ranging from the high 90th percentile to the mid 70th percentile for the 12 through 15 age group respectively. Success is measured by police/court records, regular and formal feedback from probation, schools and the family. Also, most of the curricular modules have pre and post-tests to measure retention of key content. The program is highly transportable to all communities and has even been culturally adapted to isolated Northern Native reserves. Most facilities are accessed free of charge by agreement with city corporations and boards of education and volunteers are used extensively for one-to-one assistance in program delivery.

One rather obvious and salutary application of the Attendance Model, conspicuously unavailable in the Ontario correctional armament, is the post-custodial model. This utilization of the model would provide the necessary and missing link between custody and the community. Assuming that most of the precipitating factors of an offense are neither addressed nor ameliorated by incarceration, the heightened structure and supervision of a post-release Attendance Centre would clearly provide numerous and compelling benefits. In the first instance, the Attendance Centre would provide a palpable mechanism to reduce the length of custody for young persons for whom continued committal would be of no further benefit and, coincidently, would realize this hollow provision of the *Young Offender's Act* (Section 28 (4)). This option would further alleviate the morale problem created through determinate sentencing by introducing incentives for improved progress into the custodial arena. The process enriches the post-custodial reintegrative potential, increases societal protection during a time of high-risk recidivism and would alleviate over-crowding through unnecessarily long custodial terms. This latter benefit would translate into considerable financial savings and, for that reason alone, should be appealing to correctional bureaucrats.

The third major component of the COPS program is the Community Service Order (C.S.O.) program. This program again reflects the ideals of minimal yet appropriate intervention skill building, self-esteem enhancing and community participation and responsibility. Participants, often referred by the youth courts as a separate item of disposition (Section 20 (1) (g)) or as a term of probation, are required to perform community service for charitable organizations, churches, retirement homes etc. All participants are individually and formally assessed and placements are made to reflect the strengths, interests and needs of each person. Intransigent youth may complete their community service as part of the Attendance Centre program. In these instances, difficult youth, under supervision of the regular Attendance Centre staff, will complete woodworking, welding or craft projects for later donation to charitable services. Most participants, however, fully benefit from their regular community service placement and may continue on, as voluntary or salaried workers, when their court-ordered requirements are satisfied. With more adequate funding and a more sensible interpretation of the *Young Offender's Act* the C.S.O. program could be expanded to include a Personal Service (Section 20 (1) (f)) and a Fine Option component and could assume a pre-charge/voluntary or Alternative Measures format. These latter adaptions would reduce court costs and delays (responding to the Askov decision) and would better reflect the ideals of the *Young Offender's Act*. Since the *Young Offender's Act* increased the amount a young person may be fined by the court, from $25.00 under the *Juvenile Delinquents Act* to $1,000.00, the need for Fine Option Programs, inflation notwithstanding, would appear to have increased commensurately.

This paper has argued for the establishment of a logical continuum of service for children and young persons in this province. Ironically, the apparently polarized agendas of both liberals and conservatives would be mutually served by such a comprehensive policy/program initiative. A liberal, proactive approach to crime prevention would avoid conservative, reactive notions of detention/incapacitation, preferring early assessment, diagnosis and treatment. The former notion is based on identification of risk; the latter assumes the primary importance of an assessment of need. Since both these approaches are essentially indivisible, ideological motivations aside, we would argue for the relative importance, in policy and program, of an early, liberally conceptualized needs assessment to promote optimal primary prevention efforts. Humane and efficacious treatment of needs, after all, results in a general reduction of risk. In other words, the necessiity for reactive, drastic intrusion, based on risk, is inversely proportionate to the amount of proactive, minimal intervention targetting need, presuming, of course, that services are selectively and sufficiently available to respond differentially to a rigorous risk/needs assessment. The Community Options Program, we propose, represents a flexible, cost-efficient and effective service response to an intensive risk/need assessment of children and young persons both at early risk for law-breaking and criminal behavior as well as throughout the spectrum of our formal, youth justice system. The Attendance Centre component, in particular, is sufficiently and inherently adaptable to respond to a constellation of diagnosed risk/need: familial, social, educational, cultural, psychological, sexual and spiritual issues are capably addressed through its broad, internal curriculum or its parallel and adjunctive referral mechanisms. While this service offers particular promise for early assessment and intervention with "under 12s," the international precursors of the Ontario model were reserved for persons already well enmeshed in the formal criminal justice system. In England, for instance, Attendance Centres originated in 1948 as a response to a critical institutional overcrowding. Young offenders aged 12 to 21 were ordered to attend Centres for periods of work and training during their leisure time, typically on weekends. Later legislative provisions, introduced in 1972, designated attendance as a condition of a probation order, the hours of attendance, number of days, nature of the work etc. specified by court order.

These Day Training Centres, as they are called, programmatically address broader, underlying problems such as social skills, life skills, education and employment. Similarly, Periodic Detention Work Centres were developed in Australia in 1962, reflecting a strong penchant for intensive, less expensive non-residential programming. New Zealand's adult Non-Residential Periodic Detention Program also accentuates a community-based, minimally intrusive response to crime. Weekend work projects are enriched by lifeskills/communications lectures and growth discussions held on weekday evenings. As we see, intensive, non-residential programs are remarkably and functionally adaptable to the need/risk, age and lifestyle schedules of their participants. Internationally, they have been successfully employed to reduce institutional overcrowding — a pressing problem in Canada's young offender and adult correctional systems. Demographically, of course, the problems are greatest in Canada's most populated provinces. The category between the ages of 15 to 24, a statistically crime-prone group, currently includes one in six Canadians, one out of four residing in Ontario. One in three Canadians lives in our three major cities, Montreal, Toronto or Vancouver. These demographics pose obvious logistical and financial problems (and opportunities) for crime prevention practitioners. Ominously, there has been no concerted or substantive effort to develop a uniform policy and program to address youthful urban crime and the overcrowding of youth institutions in our most populous areas. Toronto, for instance, does not have a single Attendance Centre program, having failed to support the short-lived Metro Attendance Centre Program (M.A.C.) in the mid-1980s. Paradoxically, the argument espoused by the provincial Attorney General's office, especially as it relates to their opposition to Alternative Measures, is that preventative/diversionary mechanisms "widen the correctional net" and are too costly. This position, of course, is patently nonsensical, reflecting a narrow, reactive understanding of corrections and based more on ignorance than on fact. The alternative, of course, is to continue our expansion of the carceral continuum.

Simply increasing police strength has been an unimaginative and demonstrably unsuccessful strategy at reducing rates of violent reported crime. Yet, according to Koenig (1991), Canadian police departments in 1988, on average, spent $82,334 per police officer, amounting to $169 for every Canadian. If one percent of the Metropolitan Toronto Police Department's annual gross operating budget were withheld and transferred to a municipal Attendance Centre service, four such Centres would be operational in one year. Over a five-year period, this process could potentially result in the creation of 20 Attendance Centres in the city of Toronto. Given the dubious and costly relationship between increased policing and crime prevention, such a simple rechannelling of dollars and common sense would appear to warrant serious consideration. We have argued that primary prevention initiatives, based upon social welfare and social development concepts, are ultimately and invariably less costly and more beneficial than the traditional, reactive "cops, courts and corrections" syndrome. We have clearly demonstrated that such a reorientation of present policy and practice is not foreign or contradictory to the prevailing legislation, but, rather, would usher in a desireable return to original principles, reaffirmlng primary integrity of purpose. This can achieved at no additional financial cost.

In fact, an intelligently planned and co-ordinated effort would result in a considerable reduction in current financial and social encumbrances.

The weary, unimaginative lament of government, that there are "no new dollars" available for preventive or alternative services, is both specious and tautological. Self-impoverishment is at the heart of Ontario's dominant youth Justice system. Heightened policing, expanded judiciary and increased incapacitation of offenders all conspire to perpetuate this sorry state of corrections. This malignant trinity of "cops, courts and corrections" is internally capacity driven — continuously bloating itself on its own systemic dysfunction. This 'system' somewhat senselessly parallels Parkinson's Law: work expands so as to fill the time available for its completion. The 'system,' however, is not without its own set of internal checks and balances. For instance, a proliferation of informal Alternative Measures and Youth Justice Committees would lessen the need for traditional policing and a costly and protracted court process. Review of custody mechanisms would, by truncating total days in custody, further free up dollars for more proactive services. In Ontario, for instance, the 1986-87 total number of secure facility days for phase II — or 16 and 17 year old young offenders — amounted to about 230,000 days' stay, up 72 percent over the previous year. If one third of this time were spent in postrelease, reintegrative Attendance Centres, the province would free up 76,666 days, or 10.7 million dollars annually, for more humane and preventative initiatives. (Of the 4,277 offenses committed by phase II offenders in secure custody, only 132, or 3.08 percent, were violent crimes, 3 of which, or .07 percent were homicide.) This reallocatable amount of "no new dollars" would, for example, provide funding for approximately 100 Attendance Centres in the province. This simple cost efficiency does not begin to reflect the additional dollars saved by diverting young persons from the criminal justice process as well as a reduction of an unacceptably high post-release recidivism through a planned and structured reintegration model. These modifications of the traditional correctional paradigm will also have the immediate and long-term palliative impact of moderating the current contagion of prison building that has come to dominate the past quarter century of Canadian corrections. Attendance Centres have been recommended, in Ontario as well as across Canada, by private and public sector task forces as viable options to incarceration. Clearly, however, a reduction and strict reallocation of conventional "cops, courts and corrections" dollars will be required.

The Young Lewisham Project

One final innovative scheme for dealing with delinquent youth in the community was The Young Lewisham Project (YLP) in London, England. Originally conceived in the mid-1970s its primary purpose was the prevention of youthful offending. The project offered three major program components:

(1) An auto repair and renovation "club" where young men worked on restoring old cars to certified working order. Automobiles were donated by local scrap yards, used car dealers and private citizens and garage space was similarly donated. Groups of four to six young men would work on restoration guided by an experienced volunteer mechanic. Each boy during an approximately one-year period would work on, and eventually be given the ownership of the auto. This addressed a prominent crime in England, that of "taking away and driving" or in North American terms, auto theft. Adolescents not only learned valuable mechanical skills, but it was a valuable way in which to learn values associated with co-operation, planning, and long-term goals.

(2) After school drop-ins were a second component of the program. These were run at a local community centre and included counselling, homework assistance, planned group discussion and help with familial and personal problems. The groups were targeted at boys and girls age ten-14 who were considered most at risk for involvement in crime. This is particularly true in those cases where meaningful, construc- tive recreational activities were not available.

(3) Discussion groups and neighbourhood work were the final outcroppings of the project. In the former, groups of from seven to 30 individuals would gather on a weekly basis to discuss issues such as employment, sex, legal rights, health, birth control, homosexuality, education and the police. This was done on some occasions with the participation of a guest speaker. Adolescents benefitted from the open forum and the ability to discuss issues that were often not addressed in either the home setting or the school.

The neighbourhood work group was another integrated portion of the YLP which took community prevention to youth hang-outs. Concerted efforts were made to move young people into the structure of discussion, work and drop-in groups and away from community "trouble spots" where experience had taught that many were bound to negative interactions with the community and eventually authorities.

Conclusions: Filling in Rhetoric

If the promise of the YOA and its principles are to be met, Ontario will have to undergo a pronounced shift from corrections towards the community. This will involve the consideration of viable community alternatives for young people in — or potentially in — conflict with the law. While there has been little evidence of such a shift to date, and much vacuous rhetorical posturing regarding the lack of financial wherewithal to effect change, we have attempted to demonstrate in this paper that the funds already exist for new programs. What is required is no less than a shifting of a small portion of funds currently committed to the carceral network to viable alternatives. We have presented three multi-dimensional programs which have some potential to rehumanize our approach to young offenders in Ontario. There is no need to invoke the old Marxian dictum concerning needs and assets in our society, even in the heart of "socialist" Ontario, but there is a desperate need to recognize the inherent worth of our young people and the potential for disastrous societal consequences if we continue down the path of carceral brutality.

References

Canada. Department of Justice. 1989. *The Young Offender's Act: Proposals for Amendment*. Ottawa: Supply and Services.

Canada. Department of Justice. 1991. *Report of the Workshop on Alternatives to Custody*. Ottawa: Department of Justice.

Canada. Parliament. House of Commons. 1981. *Debates*. 32nd Parliament, 1st session, 13 March-21 April 1981. Ottawa: Queen's Printer.

Canadian Centre for Justice Statistics, Statistics Canada. 1984-89. *Youth Court Statistics*. Preliminary Data. 1984-85 (December 1988). 1985-86 (December 1988). 1986-87 (April 1989). 1987-88 (April 1989). 1988-89 (August 1989).

Canadian Centre for Justice Statistics, Statistics Canada. 1990. "Sentencing in Youth Courts, 1984-85 to 1988-89." *Jusristat Service Bulletin*. 10(1) January.

Caputo, T. 1987. "The Young Offenders Act: Childrens' Rights, Childrens' Wrongs," in *Canadian Public Policy.* xiii: 2: 125-143.

Caputo, T. and D. Bracken. 1988. "Custodial Dispositions and the Young Offenders Act" in Hudson, J. *et al* (eds.), *Justice and the Young Offender in Canada.* Toronto: Wall and Thompson.

Caputo, T. 1991. "Pleasing Everybody Pleased Nobody: Changing the Juvenile Justice System in Canada" in Samuelson, L. and B. Schissel (eds.), *Criminal Justice: Sentencing Issues and Reform.* Toronto: Garamond.

Clark, B.M. 1985. *Diversion: A Less Restrictive Alternative.* Toronto: Metro Children's Advisory Group.

Clark, B.M. 1987. "Post-Attendance Release Centres: The Missing Link Between Custody and Community." Project Description to The Ontario Ministry of Community and Social Services.

Clark, B.M. 1988. *Making It or Breaking It: The Community and Alternatives.* Toronto: Ontario Social Development Council.

Clark, B.M. and D.H. Clark. 1987. "Crime Prevention Services for Indian Youth: The Native Options Program." Paper presented to the 13th Annual Canadian Indian Teacher's Education Program, Walpole Island, Ontario.

Clark, D.H. 1986. *The North Caribou Lake Band Juvenile Attendance Centre Program: A Project Proposal and Description.* Ottawa: Ministry of the Solicitor General.

Cohen, S. 1985. *Visions of Social Control.* London: Polity Press.

Doob, T. 1992. "Trends in the Use of Custodial Dispositions for Young Offenders," *The Canadian Journal of Criminology.* Vol. 34, 1: 75-84.

Ericson, *et al.* 1987. "On the Privatization of Punishment," *Canadian Review of Sociology.*

Foucault, M. 1977. *Discipline and Punish: The Birth of the Prison.* New York: Random House.

Hudson, J., J. Hornick and B. Burrows. (eds.). 1988. *Justice and the Young Offender in Canada.* Toronto: Wall and Thompson.

John Howard Society of Ontario. 1989. *Reform Bulletin.* Ontario: John Howard Society.

Jones, K. 1960. *Mental Health and Social Policy.* London: RKP.

Kenewell, J. *et al.* 1991. "Young offenders" in Barnhorst, F.R. and L. Johnson, (eds.), *The State of the Child in Ontario.* Toronto: Oxford University Press.

Koenig, D. 1991. *Do Police Cause Crime?* Ottawa: Canadian Police College.

Leschied, A. and P.G. Jaffe. 1989. "Implementing the Young Offenders Act in Ontario" in Hudson, J. *et al.* (eds.), *Justice and the Young Offender in Canada.* Toronto : Wall and Thompson.

Lewis, S. 1992. *Youth and Race Relations in Ontario.* Toronto: Government Printing.

O'Reilly-Fleming, T. 1992. "The Dark Factory: Prison Conditions, Life Imprisonment and The Politics of Release" in McCormick, K.R.E. and L. Visano (eds.), *Canadian Penology: Advanced Perspectives and Research*. Toronto: Canadian Scholars' Press Inc.

O'Reilly-Fleming, T. 1993. "Swarms, Runs and Wildings: Youth Culture, Racism and Alienation" in Schissel, B. and L. Mahood (eds.), *Social Deviance in Canada*. Toronto: Oxford University Press, forthcoming.

O'Reilly-Fleming, T. and B. Clark (eds.). 1992. *Youth Injustice: Canadian Perspectives*. Toronto: Canadian Scholars' Press Inc.

Ratner, R. and J. McMullan. 1985. "Social Control and the Rise of the Exceptional State in Britain, the United States and Canada" in Fleming, T. (ed.), *The New Criminologies in Canada: State, Crime and Control*. Toronto: Oxford University Press.

Smith-Gadacz, T. 1983. "Speak No Evil, Hear No Evil?: Juveniles and the Language of the Law" in Fleming, T. and L. Visano (eds.), *Deviant Designations: Crime, Law and Deviance in Canada*. Toronto: Butterworths.

The Young Lewisham Project. *Annual Report*. Feb. 77. London.

Turk, Austin. 1969. *Criminality and Legal Order*. Chicago: Rand McNally.

West, G. 1984. *Young Offenders and the State: A Canadian Perspective*. Toronto: Butterworths.

Legislation

An Act to Amend the Young Offenders Act, the Criminal Code, the Penitentiary Act and the Prisons and Reformatories Act (Bill C106), Statutes of Canada 1984-85-86, c.32.

Child and Family Services Act, Statutes of Ontario 1984, c.55.

Bill C-61, *Juvenile Delinquents Act*, Revised Statutes of Canada 1970, c.J-3

Bill C-192, *An Act Respecting Young Offenders and to Repeal the Juvenile Delinquents Act*, 3d Sess., 28th Parliament, 1970-71-72.

Faulty Powers: The Regulation of Carceral and Psychiatric Subjects in the 'Post-Therapeutic' Community[1]

Robert Menzies and Christopher D. Webster[2]

Introduction

In this chapter we present some preliminary findings from a six-year longitudinal study that charts the "transcarceral careers" (Lowman, Menzies and Palys, 1987; Webster and Menzies, 1987) of 162 people who were processed through the Canadian medico-legal system between 1979 and 1985. Building on earlier work that focused primarily on the dangerous behavior and institutional experiences of criminal defendants remanded for psychiatric assessment (Menzies, 1985b; Menzies, Webster and Sepejak, 1985a), we extend the analysis to consider the institutional pathways through which forensic subjects are propelled in response to their perceived dangerousness, disorder and dependency. Our concentration is on the cycles of detection, supervision and constraint that perpetuate psychiatric and criminal statuses, and on their correspondences with wider rhythms that emerged during the 1980s in the public and private ordering of mental and criminal populations.

The past two decades have witnessed a major infusion of research and writing, from a number of different and relatively autonomous fronts, on the shifting patterns of institutionalization being experienced by individuals entwined in the parallel regulatory systems of medicine, welfare and law. Theorists, researchers and practitioners in a range of disciplines have come to recognize the significance of "deinstitutionalization" as a defining feature of care and control systems in contemporary north-western societies. There is general unanimity about the scale and salience of this trend toward the systematic transfer of prisoners, mental patients, young offenders and other "delinquent and defective" cohorts from closed institutional settings into the terrain of civil society (Austin and Krisberg, 1981; Chan, 1992; Lerman, 1982; Scull, 1984).

And most acknowledge the quiet desperation of deinstitutionalized lives in a non-therapeutic or post-therapeutic community, where ex-mental patients and decarcerated people of all kinds are consigned to the backwaters of homelessness, urban slums and semi-institutionalized ghettoes of dependency, and where the very idea of "community therapy" has become an ironic oxymoron (Brown, 1985; Cohen, 1990; Dear and Wolch, 1987; Lamb, 1984; Torrey, 1988).

Yet here the consensus ends. Even a cursory summary of the literature on deinstitutionalization reveals an extraordinary divergence of accounts and prescriptions that are collectively traceable to the discordant substantive emphases, disciplinary allegiances and ideological convictions of their competing proponents.

In the mental health arena, for example, the 'downsizing' and closure of psychiatric hospitals have typically been described through the limiting discourse of clinical policy and therapeutic practice (Bachrach, 1983; *British Columbia Ministry of Health*, 1987; King, Raynes and Tizard, 1971; Talbott, 1984; Wing and Olsen, 1979). The work has been unremittingly empiricist and institutionalist in its approach. Organizational accounts and research reports are principally concerned with the general distribution of mental patients between hospitals and the streets, with the establishment and operation of "community care" facilities, and with the general mental and social adjustment of people who are discharged from the confines of traditional asylums. On the whole, investigators and authors have tended to address deinstitutionalization as an administrative or epidemiological problem manifested at the aggregate level, or alternatively as a matter of individual pathology and its treatment (Almond, 1974; Bloom, 1973; Kiesler, 1987; Olsen, 1979; Pasamanick, Scarpitti and Dinitz, 1967). What is absent from the majority of this literature, and from the writings of even the most eloquent critics of mental dehospitalization (Dear and Wolch, 1987; Lewis, 1991; Ralph, 1985; Torrey, 1988), is any sustained contextualization of the phenomenon, or a genuine reckoning with the transcendent forces that extend beyond the frontiers of mental health and hence cry out for a fully recursive, multi-systemic analysis.

These deficiencies are paralleled in the penology literature by an even more abstracted brand of empiricism that is geared to predicting the offensive potential of carceral subjects and to modelling their "criminal careers" (Blumstein, Cohen, Roth and Visher, 1986; Farrington and Tarling, 1985; Moore, Estrich, McGillis and Spelman, 1984; Nuffield, 1982). In this enterprise decarceration is treated as a scientific problem of classification which is to be resolved through the development of sophisticated sentencing formulae, actuarial tables, risk assessment scales, and sundry other devices aimed at pointing and measuring the public danger posed by diverted criminals and former prisoners. These efforts are replicated in the field of forensic psychiatry, where in recent years renewed energies have been aimed at predicting the dangerousness of mentally disordered offenders released from hospitals for the criminally insane and other settings (Ashford, 1989; Hall, 1987; Steadman, Robbins and Monahan, 1991; Thornberry and Jacoby, 1979; Toch and Adams, 1989).

In both cases, whether officials and researchers are engaged in the practice of correctional classification or medico-legal risk prediction, the systemic sources and implications of decarceration are once again generally ignored. There is little attempt to theorize the careers of criminal offenders in wider structural or cultural context, or to undertake a fully reflexive or integrative analysis that would chart these official pathways beyond the perimeters of unitary, insular organizations. As a consequence, the "community corrections" field replicates writings on "community mental health" in dissociating the private lives of subjects from the state and civil realms they inhabit, and through whose policies and discourses their activities and experiences are constructed. Moreover, these orthodoxies converge in treating individual components of the transcarceral system, whether explicitly or by default, as discrete formations with no mutual ties or effects, and as separate topics for clinical, criminological and administrative study.

A third and intensely oppositional fix on deinstitutionalization has emerged with the thinking and writing of revisionist historians and sociologists of social control. Initially energized by the sweeping "destructuring impulse" (Cohen, 1985; Pearson, 1975) that characterized various reform movements in the 1960s, new ways of conceiving deviance and its repression have surfaced, with far-reaching implications for authorities and subjects alike, and for those academics who have been endeavoring to find meaning amid the chaos of the past two decades (see Chan, 1992; Cohen, 1988, 1991; Garland, 1985; Garland and Young, 1983; Lowman, Menzies and Palys, 1987; Mathiesen, 1980). Many commentators, especially those social control theorists inspired by Foucault's vision of a post-modern, disciplinary society (where power is dispersed to peripheral locations and the state becomes only one among many actors in the penality business) have declared that the systemic transformations implicated in deinstitutionalization, decarceration, delegalization and parallel processes are immanent and even millenarian — that they represent a qualitatively different grid for organizing authority and for governing the citizenry (see Burchell, Gordon and Miller, 1991; Donzelot, 1979; Foucault, 1977, 1978; Lowman and Menzies, 1986; Miller and Rose, 1986; Rose, 1990).

Revisionist social control accounts of a deconstructed, centrifugal, panoptic, "minimum security" society (Blomberg, 1987) are evocative, compelling and hard to ignore. They provide an engaging and theoretically rich contrast to the rigid empiricism of most clinical and criminological offerings on the subject. At the same time, as Chan (1992) recently pointed out, much of the critical control literature is virtually devoid of either aggregate or ethnographic data about the deinstitutionalization phenomenon and its effects on medico-legal organizations and their human subjects. Further, this work is steeped in post-structuralist thinking, to the extent of de-centering the state and its agents into epiphenomenal status or near non-existence, or at the very least into formations that are more discursive than real in their content (see Melossi, 1990; Rose, 1990).

Formulations of this kind depict deinstitutionalization as a global, non-recursive, unidirectional flow of power and people from inside to outside, from closed to open spaces, from carceral centers to the community. But in so doing they are inclined to overdraw the magnitude and underplay the dialectics of these systemic re-alignments. The state, repressive law, penal and therapeutic institutions and public authorities of all kinds have scarcely been displaced by these dispersal movements. The effect has been far less a holistic return to "community control" practices than a layering of ptolemaic epicycles that are intricately interconnected, and that function to sling people through sequences of arcs with the continued potential for recycling back into prisons, hospitals and other contained sites of centralized penal authority.

What revisionist social control theorists need, in short, is a body of grounded institutional and ethnographic data, yielded from multiple sites and sources, that would chart and evaluate the structural and human results of deinstitutionalization practices. This in turn requires a rapprochement and fusion between the empiricist "community adjustment" and "criminal careers" projects of clinicians and criminologists cited above, and the theoretically charged structural analyses of critical socio-legal scholars. Work of this kind has to mediate between the competing dangers of hypostatizing or celebrating formal control institutions on the one hand, and dissolving them into the shadowy half-life of a post-structuralist netherworld on the other. Moreover, analysis needs to be multi-disciplinary in organization, supra-institutional in scope, and iteratively pointing to both the substantive transformations in control systems and community, and the resulting diversions and reversals in the life courses of individuals who are caught up in the workings as either officials or subjects.

What we present here is barely a beginning. A few tables charting the six-year "medico-legal careers" of 162 forensic patients from one Canadian city will not, of course, remotely answer or resolve any of these acute questions about the changing relationships between carceral and community modes of social regulation, or between the deviant citizen and the Canadian state. Nor will they very far advance our collective knowledge of deinstitutionalization policies or their long-term effects on penal and psychiatric populations.

Nonetheless, detailed longitudinal data such as those depicted below — which follow carceral and therapeutic subjects through multiple institutional locations and into the community over an extended time period — can offer an especially transparent window on the intermeshed mechanics of legal and medical ordering, and their relation to allied stations in both inclusionary and exclusionary locations. They also may provide a preliminary glimmer of understanding about the relative power of state institutions to maintain regulatory ownership of their conscript clientele (Friedenberg, 1976), and about the position of prisons and hospitals in the complex infrastructure of transinstitutional systems. Forensic patients like the METFORS remand subjects described here, by their very status as dual denizens of criminal and psychiatric terrains, are the consummate guides for such an expedition. By following their tracks over the course of their 72-month itinerary, we may be able to establish a fresh vantage point for mapping the institutional and civil sites they visit, and for theorizing their operations.

Context, Subjects and Method

This study follows a cohort of 162 Canadian medico-legal subjects through six years of their circulation within a constellation of carceral institutions, mental health organizations and various realms of the community. The tracking of these people began with their arrest and subsequent psychiatric remand, between March and June of 1979, to the Metropolitan Toronto Forensic Service. METFORS — a pre-trial assessment center[3] located on the grounds of the Queen Street Mental Health Centre on the west side of Toronto — has been the subject of much research and controversy over the 15 years that have passed since its inception in 1977. Established as a multi-disciplinary[4] court clinic specializing in the determination of fitness to stand trial, METFORS has been the lightning rod for a persistent stream of legal, academic and clinical debate over a range of issues, including the legal protection of accused persons suspected of mental disorder, the accuracy and efficacy of forensic decisionmaking, the capacity of clinicians to predict the dangerousness of their patients, and the delegation of sentencing authority from criminal court judges to mental health professionals (see Butler and Turner, 1980; Chunn and Menzies, 1990; Menzies, 1989; Menzies, Webster and Sepejak, 1985a,b; Webster, Menzies and Jackson, 1982; Webster and Menzies, 1989).

The subjects of METFORS assessments, like forensic patients elsewhere (Gibbens, Soothill and Pope, 1977; Hodgins, 1992; Teplin, 1984; Toch and Adams, 1989; Steadman, 1979), comprise a unique source of understanding about the relations between penal institutions, therapeutic settings and civil society. As candidates for both criminal sanctions and psychiatric interventions, "mentally disordered offenders" frequently become the targets of multiple regulatory practices aimed to both punish their infractions and restore them to normalcy. Their carceral careers can follow intricate, and often elliptical, orbits across public and private domains, among courts, prisons, hospitals and clinics, into the streets and through a blinding array of halfway and hybrid sites of containment and supervision (McMain, Webster and Menzies, 1989; Menzies, 1987). Whereas the more "typically criminal" among them will be cordoned off for penal confinement, and the "overtly mad" consigned to medical contexts, there is a core group — ranging from one-third up to one-half of all forensic assessment patients — who will continue to cross the frontiers of judicial, penal and mental settings with frequency and regularity.

The pathways carved out by such persons, and their institutionally ascribed conduct and reasons for intervention at each stage, hold an enormous heuristic potential for bringing the systems themselves, and their mutual frictions and attractions, into relief. Such a "subject-centered" analysis, notwithstanding the obvious limitations of official data and aggregate statistical summaries, can at the very least demonstrate some of the cumulative human effects of mega-institutional control practices, and help to assess the workings in concert of multiple systems and agencies. These observations in turn might shed some light on the phenomenon of deinstitutionalization as it is experienced with the passage of time by one Canadian cohort of medico-legal subjects.

This was clearly an appropriate group with which to work. The METFORS subjects entered the front end of the project as highly institutionalized, criminalized, and pathologized women and men. The attributes of these 162 people were very much like those evident among forensic populations previously described in a legion of studies about psychiatric and psychological remands (see above). They were typically marginal individuals, predominantly young unmarried men with substantial histories of psychiatric and carceral contacts. More than 85 percent were male, and 60 percent were under the age of 30. Only 17 subjects (one in ten) were cohabiting at the time of their arrest, 53 (two in five) were unemployed, whereas only five (2.1 percent) had any university education. More than half (52.1 percent) had been previously hospitalized for a mental disorder; one-third (34.1 percent) had attempted suicide; nearly three-quarters (111 of 152 for whom data were available) had a prior criminal record; and more than two of every five (43.5 percent) had spent time in prison. Fifty-five people (34.2 percent) went to METFORS facing a property-related charge, 32 (19.9 percent) were awaiting trial, respectively, for violent offenses, and infractions related to alcohol or other drugs, and 26 (16.1 percent) were charged with "technical"[5] violations.

In their initial brief assessment at METFORS, 42.8 percent of the defendants (68 of 161) were considered mentally disordered by the professional classifiers, and 21.2 percent were deemed either unfit to stand trial (N = 14) or questionably fit (N = 19). One-third each (N = 52 and 59) were evaluated to be in need, respectively, of a custodial setting or further forensic assessment; one-quarter (N = 33) were recommended for inpatient mental treatment; and one-half (N = 81) were identified as candidates for outpatient care. Across the entire sample, 22 (14.5 percent) were characterized by the presiding psychiatrist as definitely dangerous, 116 (76.3 percent) as potentially dangerous, and only 14 (9.2 percent) as not at all dangerous. Among the 154 people for whom sentencing data were available, 65 (42.2 percent) were initially sentenced to prison following their METFORS assessment, 45 (29.2 percent) received probation, and 44 (28.6 percent) were found not guilty or had their charges withdrawn (see Menzies, Webster, McMain, Staley and Scaglione, in press).

In other projects we have undertaken a more detailed examination of the forensic remand process itself (Webster, Menzies and Jackson, 1982), and we have developed some extensive micro-level analysis of decision-making practices among clinical classifiers and their implications for criminal defendants (Menzies, 1987; Webster and Menzies, 1987; Webster and Menzies, 1989). In what follows here, we focus our attention instead on the subsequent conduct, and the patterns of de-re-trans-institutionalization that defined the forensic careers of subjects during the 72 months following their initial METFORS assessment.

Carceral, Therapeutic and Community Follow-up

In earlier research at METFORS (see Menzies, 1987, 1989; Menzies, Webster and Sepejak, 1985a,b; Webster and Menzies, 1987) we had developed a two-year follow-up of mentally disordered offenders, primarily for the purpose of assessing the capacity of clinical decision-makers and semi-structured psychometric instruments to predict the future dangerousness and violence of forensic subjects. This work was based on a 24-month review, following the initial remand for each individual, of Ontario Correctional Services printouts,

Canadian Police Information Centre (CPIC) files, psychiatric hospital records for METFORS and six other institutions within a 100-mile radius of Toronto, and provincial death registries.

While these materials generated some interesting findings on the relationship between forensic judgments and subsequent violent behavior among this sample, they were not originally designed to chart in detail the complex institutional pathways through which these people were dispersed following their initial forensic experience. The outcome data were also temporally limited, with no systematic information available outside of the two-year research frame. Moreover, the confinement of correctional documents in the first study to the province of Ontario resulted in some major omissions, given the frequently picaresque migratory patterns of these medico-legal subjects.

Accordingly, over the past several years we been working to compile a more extensive and representative matrix of officially registered data on the 162 criminal defendants described above. In addition to background sociodemographic and medico-legal variables, information was collected on a variety of judgments rendered by METFORS clinicians and trained coders on subjects' mental status, criminal attributes, institutional prospects and propensity for dangerousness to self and others. Most directly pertinent to this chapter, a detailed search was conducted through all available carceral and therapeutic records for a period of six years following their original 1979 METFORS remand. The data sources included:

(1)   CPIC printouts, documenting the nature, location and date of subsequent criminal charges across Canada, along with related criminal court sentences;

(2)   correctional services files, including dates of admission and release, along with prison misconducts incurred during confinement, from seven of the nine provinces in which sample members were imprisoned over the course of the six years;[6]

(3)   National Parole files for the 43 people who served federal time for criminal sentences of two years or longer, including information on prison transfers and official misconducts during confinement;

(4)   Correctional Services Canada computer data on the same group of federal inmates;

(5)   medical records from ten major psychiatric and forensic institutions in Ontario,[7] plus five general hospitals in the Toronto area; and

(6)   Ontario Death registry data banks, to identify individuals who died during the 72 months of the follow-up.

These various documents were located for each subject by a team of research assistants, who transferred the source institutional materials onto standardized forms designed to organize the follow-up data into comparable categories in judicial, penal, hospital and other settings. Descriptions of subject experience for each correctional and psychiatric contact were compiled, and original accounts of conduct, transactions, incidents (disruptive, criminal, violent and self-injurious) were transcribed in the original language (with identifying references deleted) of the organization and agency authors. The nature and effects of treatment initiatives, punishment and other interventions were recorded. In the case of criminal records, dates, locations and outcomes of all charges were noted, and dispositions cross-checked between federal and provincial sources.

The data originating with these multiple sources were subsequently aggregated into discrete files for each of the 162 subjects and chronologically ordered into time lines that charted institutional or community location, and officially recorded incidents, across the full six-year period. A coding manual was then drafted, pre-tested and revised, with the final version comprising 5040 variables which focused, in addition to subject attributes and clinical decision-making, most particularly on the myriad pathways pursued, behaviors rendered and institutions encountered by the 162 people in the wake of their initial forensic experience. Detailed summaries of each correctional and psychiatric contact were encoded, with specifications of institutional entry, transfer and exit, judicial, penal and psychiatric decisions, recorded conduct and misconduct, along with the official response of authorities in each instance, and subject status prior to, during and subsequent to each individual contact. Year-by-year summaries were also generated, which concentrated on establishing, by annual tallies, the varying statuses and medico-legal careers of each cohort member, along with all criminal, psychiatric and other incidents registered in hospitals, in provincial and federal prisons, and on the streets. The entire process of recovering the data, organizing the files, drafting the manual, coding, keypunching and computer analyzing the resulting 18,144-record data set, took up five years between 1986 and 1991.

Against the context of our earlier discussion concerning the implications of such data for critical theorizing about transformations in contemporary carceral systems, this chapter offers a preliminary glimpse at the patterns of criminality, psychiatric intervention, institutionalization, release and re-circulation exhibited by these forensic subjects over the course of this six-year chronicle. More specifically, we enlist the year-by-year follow-up summaries, extracted from aggregate statistical analysis of subject profiles, both to review their carceral, clinical and community involvements, and to summarize their officially depicted conduct over time and across a variety of organizational and open contexts.

Patterns of Penal and Psychiatric Involvement

The most prominent feature of these longitudinal forensic profiles was the frequency and intensity of institutional recurrence in the lives of medico-legal subjects. Table 1 presents an annual synopsis and aggregation of correctional and psychiatric contacts among the 162 subjects, including brief forensic assessments, forensic inpatient assessments, general psychiatric inpatient admissions, outpatient referrals, daycare or drop-in referrals, probation sentences, provincial prison sentences and federal prison sentences. Altogether there were 1334 such contacts over the six-year period (Mean = 8.39 per subject), including 711 mental health interventions (Mean = 4.47) and 623 penal encounters (Mean = 3.92). The most common form of involvement was the provincial prison sentence (N = 398), followed in order by psychiatric inpatient admissions, daycare or drop-in referrals and probation orders (N = 199, 189 and 178 respectively).

There is evidence that the frequency and amplitude of medico-legal intercessions did diminish over time for these people. For example, the average number of psychiatric contacts per subject declined from 1.49 during the first post-METFORS year, to 0.58 by the sixth annual tabulation. Similarly, imprisonments and probation terms dropped over the time span from a mean of 2.92 in Year One down to 0.92 in Year Six.

Nonetheless, the sheer magnitude of these statistics is quite remarkable. The persistence of formal institutions as dominant features in the lives of forensic subjects, who dwell for lengthy periods under the twin peaks of psychiatry and criminal law, is graphically revealed in Table 1. And the prototypical forensic career consists not of unitary, exclusionary enclosures in single institutions, but rather in sporadic or periodic cycles of intervention that accumulate rapidly in the years following the precipitating criminal contact.

Table 1. Contacts with Correctional and Psychiatric Institutions During Six-Year Follow-up													
	Year one		Year two		Year three		Year four		Year five		Year six		Total	
	N	Mean	N	Mean	N	Mean	N	Mean	N	Mean	N	Mean	N	Mean
Brief Forensic Assessments	31	0.20	24	0.15	7	0.14	7	0.04	5	0.03	8	0.05	81	0.51
Forensic Inpat Assessments	77	0.48	18	0.11	13	0.08	2	0.01	5	0.03	7	0.04	122	0.77
Inpatient Admissions	46	0.29	50	0.31	20	0.13	24	0.15	28	0.18	31	0.20	199	1.25
Outpatient Referrals	31	0.20	18	0.11	27	0.17	19	0.12	10	0.06	14	0.09	119	0.75
Daycare-Dropin Referrals	52	0.32	30	0.19	29	0.18	27	0.16	18	0.11	33	0.21	189	1.19
Total Psychiatric Contacts	237	1.49	140	0.88	96	0.60	79	0.50	66	0.42	93	0.58	711	4.47
Probation Sentences	88	0.55	25	0.16	21	0.13	16	0.10	14	0.03	14	0.09	178	1.12
Provincial Prison Sentences	129	0.81	70	0.44	64	0.40	60	0.38	41	0.26	34	0.21	398	2.50
Federal Prison Sentences	10	0.06	13	0.08	11	0.07	3	0.02	4	0.03	6	0.04	47	0.30
Total Carceral Contacts	227	1.43	108	0.68	96	0.60	79	0.50	59	0.37	54	0.34	623	3.92
Ttl Psych + Carcrl Contacts	464	2.92	248	1.56	192	1.21	158	0.99	125	0.79	147	0.92	1334	8.39

Table 2. Number of Subjects, and Mean Number of Days, Spent Various Conditions During Six-Year Follow-up

	Year one		Year two		Year three		Year four		Year five		Year six		Total	
	S	MD	S	MD	S	MD	S	MD	S	MD	S	MD	S	MD
Pre-Trial Lockup	137	32	39	10	31	6	23	6	27	5	24	4	145	63
Post-Sentence Prison	84	63	61	50	59	54	52	48	44	39	41	33	111	287
Parole-Probation-Bail	120	177	113	169	80	123	61	73	50	64	38	50	140	656
Psychiatric Inpatient	79	32	35	19	18	10	19	12	17	10	22	8	91	91
Psychiatric Outpatient	14	2	16	11	13	14	18	13	13	9	15	13	43	62
At Liberty	45	57	65	106	92	156	119	211	120	234	128	250	149	1014
Dead	0	0	0	0	0	0	1	<1	2	3	3	5	4	9
Don't Know	4	2	0	0	1	<1	1	<1	1	1	1	2	7	6

S = Number of subjects experiencing the condition. Conditions for each year are coded redundantly, and more than one may apply per subject.
MD = Mean number of days spent by subjects under the condition.

Table 2 illustrates the institutional careers of forensic subjects from a different vantage point, namely the total number of individuals, and the average annual number of days spent, under each correctional, therapeutic and community condition during the six-year follow-up. This information further reinforces the general inclination displayed in Table 1 for a high proportion of individuals to fall within the dual fields of criminal justice and mental health, and for their institutional careers to be typically characterized by recurrent and cyclical encounters, each of relatively brief duration, with both control systems and community settings.

The rightmost column in Table 2, collapsing counts across the entire six-year period, is illustrative. Almost everyone in the cohort was exposed to some form of institutional confinement, but conversely virtually no one was denied at least some time in the community, either at full liberty or under penal or therapeutic supervision. Among the 162 former METFORS remandees, 145 (89.5 percent) were locked up prior to trial at some juncture during the 72 months, 111 (68.5 percent) were subjected to a prison term, 140 (84.8 percent) experienced intermediate carceral interventions (parole, probation and bail), 91 (56.2 percent) were psychiatric inpatients, and 43 (26.5 percent) were outpatients. At the same time, 149, or 92.0 percent, among the cohort were living without any legal or medical restrictions for at least some period across the time span.[8]

The high incidence of institutional interventions illustrated in Table 2 juxtapose, however, against a general drift over time from exclusionary (institutional) to inclusionary (community) modes of response (see Cohen, 1985, 1988). Whereas intermediate interventions dominated the first two years of patients' post-METFORS experiences (with a Mean of 177 days of parole, probation or bail in Year One, and 169 days in Year Two), by the third annual tally "at liberty in the community" assumed the highest rank. By the final year of the follow-up, the "average subject" spent 250 out of the 365 days without any form of supervision or restraint, plus 50 days of intermediate surveillance. When tallied across the six years, the average number of days spent by subjects at full liberty totalled 1014 (or 46.3 percent of the total), compared with 656 days under intermediate supervision (30.0 percent), 287 days in prison (13.1 percent), 91 days as a psychiatric inpatient (4.2 percent), 62 as an outpatient (2.8 percent), and 63 in pre-trial lockup (2.9 percent). Stated differently, measures of confinement duration, as opposed to raw periodicity numbers, were associated with relatively lower indices of institutional experience. This finding accords with the longstanding observation in the mental health field that psychiatric deinstitutionalization has resulted in shorter average lengths of

Table 3. Violent, Suicidal and Other Incidents During Six-Year Follow-up

	Year one		Year two		Year three		Year four		Year five		Year six		Total	
	S	MD	S	MD	S	MD	S	MD	S	MD	S	MD	S	MD
Criminal Charges	352	2.21	254	1.60	253	1.59	250	1.57	207	1.30	160	1.01	1476	9.28
Violent Charges	47	0.30	38	0.24	26	0.16	27	0.17	22	0.14	20	0.13	180	1.13
Community Incidents Other	136	0.86	68	0.43	62	0.39	31	0.20	118	0.74	49	0.31	464	2.92
Violence-Threats	20	0.12	13	0.08	9	0.06	9	0.06	13	0.08	10	0.06	74	0.47
Violence-Actions	18	0.11	6	0.04	22	0.14	3	0.02	3	0.02	9	0.06	61	0.38
Suicide-Threats	19	0.12	17	0.11	7	0.04	8	0.05	5	0.03	10	0.06	66	0.42
Suicide-Attempts	15	0.09	7	0.04	6	0.04	1	0.01	2	0.01	3	0.02	34	0.21
Inpatient Incidents	388	2.44	159	1.00	55	0.35	83	0.52	19	0.12	39	0.25	743	4.67
Violence-Threats	175	1.10	63	0.40	25	0.16	44	0.28	10	0.06	20	0.13	337	2.12
Violence-Actions	77	0.48	27	0.17	5	0.03	16	0.10	0	0.00	5	0.03	130	0.82
Suicide-Threats	21	0.13	8	0.05	13	0.08	0	0.00	0	0.00	4	0.03	46	0.29
Suicide-Attempts	13	0.08	9	0.06	2	0.01	0	0.00	0	0.00	1	0.01	25	0.16
Outpatient Incidents	29	0.18	18	0.11	7	0.04	97	0.61	4	0.03	7	0.04	162	1.02
Violence-Threats	6	0.04	11	0.07	3	0.02	3	0.02	3	0.02	5	0.03	31	0.19
Violence-Actions	1	0.01	0	0.00	2	0.01	1	0.01	0	0.00	0	0.00	4	0.03
Suicide-Threats	9	0.06	5	0.03	1	0.01	1	0.01	1	0.01	0	0.00	17	0.11
Suicide-Attempts	5	0.03	0	0.00	0	0.00	0	0.00	0	0.00	0	0.00	5	0.03
Pre-Trial Jail Incidents	38	0.24	7	0.04	18	0.11	1	0.01	4	0.03	1	0.01	69	0.43
Violence-Threats	0	0.00	2	0.01	0	0.00	0	0.00	0	0.00	0	0.00	2	0.01
Violence-Actions	3	0.02	1	0.01	0	0.00	0	0.00	1	0.00	1	0.01	6	0.04
Prison Misconducts	62	0.39	51	0.32	92	0.58	61	0.38	41	0.26	24	0.15	331	2.08
Violence-Threats	3	0.02	2	0.01	9	0.06	1	0.01	0	0.00	0	0.00	15	0.09
Violence-Actions	12	0.08	5	0.03	23	0.14	12	0.08	11	0.07	5	0.03	68	0.43
Total Incidents	1005	6.32	557	3.50	487	3.06	523	3.29	393	2.47	280	1.76	3245	20.41
Total Violence-Threats	204	1.28	91	0.57	46	0.29	57	0.36	26	0.16	35	0.22	459	2.89
Total Violence-Actions	158	0.99	77	0.48	78	0.49	59	0.37	37	0.23	40	0.25	449	2.82
Total Suicide-Threats	49	0.31	30	0.19	21	0.13	9	0.06	6	0.04	14	0.09	129	0.81
Total Suicide-Attempts	33	0.21	16	0.10	8	0.05	1	0.01	2	0.01	4	0.03	64	0.40

hospitalization for the mentally disordered, but higher frequencies of contact, thereby ironically expanding the pool of potential therapeutic subjects (see Bassuk and Gerson, 1978; Brown, 1985; Curtis, 1986; Scull, 1984).

Further, notwithstanding the obviously prominent position occupied by the community and by mediating agencies in the inter-institutional lives of forensic subjects, it would be an error to discount the continued defining role played by the arterial centers of penal and therapeutic power in constructing their careers. The very finding that two-thirds of the METFORS defendants (111 of 162) went to prison within six years of their remand, and that more than half (N = 91) were hospitalized as mental patients, underscores the enduring and imposing presence of exclusionary institutions as magnetic polarities which maintain hierarchy and structure in the face of otherwise decarcerating, dispersing and deconstructionist forces. The terms of confinement may decline with the passage of time — and in longitudinal view the time spent behind bars and walls may not compare with that of intermediate supervision or community release — but it is the prison and hospital that remain the energizing sources of carceral power, and that make these aggregate profiles comprehensible.

In a country like Canada, with one of the highest incarceration rates in the world (Mandel, 1991) and with such a massive public investment in institutional mental care, talk of a minimum security society, where the formal exclusionary institutions are fully decentered or displaced, represents a gross misreading of aggregate correctional and medical statisics, and (when they are made available, as in this chapter) of concentrated, subject-centered longitudinal data from penal and psychiatric sources. As we suggest below, the reality — at least as viewed from the perspective of those towards whom regulatory practices are directed — falls somewhere between one-sided deconstructionist accounts of a dissipated non-system, and traditional versions of a state-centered monolithic enterprise. When the careers of medico-legal subjects are charted over time, and the operations of control and care institutions are explored, a far more multi-dimensional, dialectical, open-ended, historically and contextually specific understanding of carceral, therapeutic and community control begins to emerge.

Criminal and Other Incidents During Follow-Up

The second objective in our analysis of correctional and mental health data was to survey the officially registered conduct exhibited by the 162 forensic subjects during their time spent in prisons, hospitals and the community over the course of the follow-up period. In this section, we provide a general profile of general, violent and suicidal incidents recorded for the METFORS cohort across time and contexts, we offer some tentative explanations for the patterns of criminal and other forms of behavior that emerged in various settings, and we attempt to draw these together with our observations on the institutional dimensions of subject experiences more generally.

Table 3 presents a comprehensive year-by-year inventory of incidents charted in various institutional records against the 162 forensic subjects during their 72 post-METFORS months. These are categorized in rows according to the location of the transaction (community, hospital or prison) and the type of incident (general incidents,[9] acts of violence, threats of violence, suicide attempts and suicide threats). Totals are calculated across all six years in the two rightmost columns, and across the various contexts in the bottom five rows.

In aggregated totality, these people were involved in an extraordinary tally of officially recognized misconduct that rivalled, or even surpassed, their frequency and magnitude of institutional contacts as reviewed in the above section. Within the span of six years, the 162 subjects had registered 3245 separate incidents, for an average of 20.41 per individual. These included 459 violence threats (Mean = 2.89), 449 violent transactions (Mean = 2.82), 129 suicide threats (Mean = 0.81) and 64 overt suicide attempts (Mean = 0.40). Among the general incidents, 1940 (59.8 percent) transpired in the community, with 1476 (or 76.1 percent) of these resulting in criminal charges. In contrast, 905 (27.9 percent) of incidents took place in hospitals, and 400 (12.3 percent) in jails or prisons. Acts of violence were also most frequent on the streets (N = 241, with 180 of these leading to criminal charges), followed by violence in psychiatric institutions (N = 134) and in prisons (N = 74). According to these raw statistical profiles, even for such a highly institutionalized cohort as these forensic subjects represent, the large majority of their officially registered transgressions, criminality and violent misconducts occurred not within the confines of jails or hospitals, but in the streets.

The second major trend evident in Table 3 echoes the tendency, demonstrated earlier in the tracking of institutional pathways, for official contacts with carceral and psychiatric authorities to decline with the passage of time. Table 2 had sketched the incremental attrition of forensic subjects from the dual gravitational fields of law and medicine from Year One to Year Six of the follow-up. Here we witness the behavioral manifestation of this drift, as rates of every conduct type, in every setting, diminish as one moves further away in time and space from the original METFORS "big bang" in February through May of 1979.

Table 4. Number and Type of Criminal Charges During Six-Year Follow-up

	Year one		Year two		Year three		Year four		Year five		Year six		Total	
	N	Mean	N	Mean	N	Mean	N	Mean	N	Mean	N	Mean	N	Mean
Property	150	0.94	113	0.71	100	0.63	89	0.56	73	0.46	47	0.30	572	3.60
Violent	47	0.30	38	0.24	26	0.16	27	0.17	22	0.14	20	0.13	180	1.13
Technical-Administative	76	0.48	63	0.40	66	0.42	63	0.40	47	0.30	42	0.26	357	2.25
Drugs-Alcohol	9	0.06	21	0.13	32	0.20	44	0.28	30	0.19	34	0.21	170	1.07
Public Order	54	0.34	6	0.04	19	0.12	12	0.08	5	0.03	6	0.04	102	0.64
Nonviolent Sex	2	0.01	2	0.01	1	0.01	2	0.01	0	0.00	1	0.01	8	0.05
Parole Violation	0	0.00	3	0.02	5	0.03	2	0.01	2	0.01	5	0.03	17	0.11
Highway Traffic	14	0.09	8	0.05	4	0.03	11	0.07	28	0.18	5	0.03	70	0.44
Total Contacts[a]	167	1.05	123	0.77	159	1.00	104	0.65	77	0.48	84	0.53	714	4.49

[a] Number of separate contacts per year. Multiple charge types may appear for each contact.

This attrition effect is also evident in Table 4, which disaggregates criminal charges incurred by the cohort over the six-year period. Whatever their relative levels of incidence — with property offenses being most commonplace (N = 572), followed in order by technical-administrative violations[10] (N = 357), violent crimes (N = 180), drug and alcohol violations (N = 170) and public order offenses (N = 102) — there was again a general downward curve in the crime rate for these subjects from Year One to Year Six. The only minor exceptions were for parole violations, and drug- and alcohol-related infractions, where the somewhat higher incidence levels in later years might be partially explained by elevated opportunity.

When the results of Tables 2, 3 and 4 are combined, the general attrition of both institutional contacts, and all categories of officially registered conduct, may have some interesting implications for theorizing the relations between control institutions and the community in the Canadian context. At least for this one cohort of carceral subjects who have managed to absorb both penal and psychiatric identities, some form of genuine deinstitutionalization or decontrol process seems to be in effect.

The medico-legal apparatus is not a hermetically sealed, hypostatized leviathan. The forced migrations of people under its purview are not only internal or lateral, and more than just institutional displacement is in effect. If the temporal view is sufficiently long, and the tracking process succeeds in involving all forms of available correctional and clinical data, important centripetal forces ultimately come into focus. It becomes possible to distinguish openings in the control apparatus, and to glimpse the trails left by people who trickle out of the system altogether and for good. However institution-centered the carceral apparatus may be, accounts of a seamless, omnipresent system, permeating all reaches of state and civil society, are belied by empirical work such as this. As usual, the reality is far more complex than either the hypostatizers or the deconstructionists would have us believe. The relations between the state's control institutions and civil society are recursive, fractured, contradictory, in perpetual motion and replete with opportunities for exclusion, escape, resistance and perhaps even transformation.

Table 5. Annual and Total Number of Incidents per Subject/Year During Follow-up

	Year one		Year two		Year three		Year four		Year five		Year six		Total	
	N	Per Pt/Yr[a]	N	Per Pt/Yr	N	Per Pt/Yr	N	Per Pt/Yr	N	Per Pt/Yr	N	Per Pt/Yr	N	Per Pt/Yr
Community														
Total Months	1228		1476		1501		1530		1582		1614		8931	
Total Incidents	517	5.05	340	2.76	322	2.57	378	2.96	329	2.50	216	1.61	2102	2.82
(Criminal)	352	3.44	254	2.07	253	2.02	250	1.96	207	1.57	160	1.19	1476	1.98
(Other)	165	1.61	86	0.70	69	0.55	128	1.00	122	0.93	56	0.42	626	0.84
Violence: Acts	66	0.64	44	0.36	51	0.41	31	0.24	25	0.19	35	0.26	252	0.34
(Criminal)	47	0.46	38	0.31	26	0.21	27	0.21	22	0.17	20	0.15	180	0.24
(Other)	19	0.19	6	0.05	25	0.20	4	0.03	3	0.03	15	0.11	72	0.10
Violence: Threats	26	0.25	24	0.20	12	0.10	12	0.09	16	0.12	15	0.11	105	0.14
Suicide: Attempts	20	0.20	7	0.06	6	0.05	1	0.01	2	0.02	3	0.02	39	0.05
Suicide: Threats	28	0.27	22	0.18	8	0.06	9	0.07	6	0.05	10	0.07	83	0.11
Prison														
Total Months	503		303		303		275		217		183		1784	
Total Incidents	100	2.39	58	2.30	111	4.40	62	2.71	45	2.49	25	1.64	401	2.70
Violence: Acts	15	0.36	6	0.24	23	0.91	12	0.52	12	0.66	6	0.39	74	0.50
Violence: Threats	3	0.07	4	0.16	9	0.36	1	0.04	0	0.00	0	0.00	17	0.11
Hospital														
Total Months	177		94		46		55		47		43		462	
Total Incidents	388	26.31	159	20.30	55	14.35	83	18.11	19	4.85	39	10.88	743	19.30
Violence: Acts	77	5.22	27	3.45	5	1.30	16	3.49	0	0.00	5	1.40	130	3.38
Violence: Threats	175	11.86	63	8.04	25	6.52	44	9.60	10	2.55	20	5.58	337	8.75
Suicide: Attempts	13	0.88	9	1.15	2	0.52	0	0.00	0	0.00	1	0.28	25	0.65
Suicide: Threats	21	1.42	8	1.02	13	3.39	0	0.00	0	0.00	4	1.12	46	1.19
All conditions														
Total Months	1908[c]		1873		1850		1860		1846		1840		11177	
Total Incidents	1005	6.32	557	3.57	488	3.17	523	3.37	393	2.55	280	1.83	3246	3.49
Violence: Acts	158	0.99	77	0.49	79	0.51	59	0.38	47	0.31	46	0.30	456	0.49
Violence: Threats	204	1.28	91	0.58	46	0.30	57	0.37	26	0.17	35	0.23	459	0.49
Suicide: Attempts	33	0.21	16	0.10	8	0.05	1	0.01	2[b]	0.01	4[b]	0.03	64	0.07
Suicide: Threats	49	0.31	30	0.19	21	0.14	9	0.06	6	0.04	14	0.09	129	0.14

[a] Incidents per patient/year under each condition (number of incidents divided by number of patient-months per condition times 12). [b] One successful suicide attempt. [c] Out of a total of 1908 months per year (159 subjects x 12), and 11,448 months altogether. Remaining patient-months are unaccounted for.

Table 5 focuses yet another lens on the six-year carceral careers of these 162 former mental remandees. Here we consider the various incidents of disruption, criminality, violence and self-abuse recorded against subjects as a direct function of time and institutional or community location. For each of the six annual periods, and for all conditions and forms of conduct, incidents are displayed both in raw form and as an expression of frequency per subject/year.[11] Total raw and subject/year rates are provided across all six years in the right-hand columns, and across all conditions (community, prison and hospital) in the bottom set of rows.

There were dramatic differentials in absolute and relative rates across both time and space. The single most conspicuous feature of Table 5 is the emergence of psychiatric institutions as by far the most facilitating contexts for the registration of all forms of problematic conduct. Across the entire 72-month time frame, for each subject/year of mental hospital confinement, there were 19.30 recorded incidents, 3.38 acts of violence, 8.75 violent threats, 0.65 suicide attempts and 1.19 suicide threats. These compared with relative rates of 2.82 general incidents, 0.34 acts of violence, 0.14 threats of violence, 0.05 suicide attempts and 0.11 suicide threats per subject/year in the community; and corresponding quotients of 2.70 (general incidents), 0.50 (violent acts) and 0.11 (threats of violence) in prison. Whereas community and penal conduct incidence levels were relatively concordant, the psychiatric institutions were obvious outliers. The intensity of official monitoring in mental settings, the proactive use of progress reports and other forms of documentation, and the high concentration of troubled people in congested and emotionally charged environments, may all have contributed to these disproportionate institutional outcomes.

The trends exhibited in Table 5 also lend further perspective to the attrition phenomenon described earlier. In both prison and hospital, the ex-METFORS defendants demonstrated lower relative incidence rates toward the later years of the follow-up. For example, for every 12 months of subject imprisonment, cohort members registered 2.39 total misconducts in Year One, and 1.64 in Year Six (a decrease of 31 percent). Similarly, the total psychiatric hospital incident rate (again, per subject/year) was 26.31 in Year One and 10.88 in Year Six (this time, a decline of 57 percent).

However, the most dramatic decrements in relative subject susceptibility to official detection occurred in the community itself. Here the general incident rate fell a full 68 percent (from 5.05 to 1.61 per subject/year); there were corresponding drops, between the first and sixth years, in the proportionate community rates of criminal charges (65 percent, from 3.44 to 1.19) and of violent transactions (59 percent, from 0.64 to 0.26). Clearly, this progressive decline in detectability, especially evident in community settings, coupled with the 31 percent increase in time spent on the streets by cohort members from the first to sixth years, combined at least partially to explain the lower frequencies of misconducts, crimes and violence registered against subjects with the passage of time. While data beyond the six-year research frame were unavailable, there was no reason to believe that this attritional trend would not continue as successively more members of the post-METFORS cohort managed to elude the twin forces of medicine and law.

Discussion and Conclusions

As we remarked above, this longitudinal survey of 162 Canadian criminal defendants can provide at best a mere fleeting and wide-angle view on the state and civil forces that constituted their six-year "forensic careers," and on the mutual arrangements between various regulatory institutions and practices that shaped their experience and apprehended conduct. Given the aggregate form of our official data, we can do little more than speculate about the ideologies and interests informing the decisions of individual carceral and clinical authorities. Similarly, the inter-subjective responses of medico-legal subjects to their penal and therapeutic regulation cannot be addressed. What we have endeavored to construct, instead, is mainly a frame and flowchart — the first to enlist such detailed and integrated data across a variety of institutional, community, legal and medical sites — in order to direct subsequent work with more qualitative and idiographic emphases.

But on their own terms as well — and all the more so given current debates among revisionist control theorists about the "destructuring impulse," about the attributes and pathways of carceral power, and about the relationship between state institutions and the community — data such as these may perform a function. They can offer a competing source of knowledge, particularly when public and professional debate about punishment and treatment is so frequently either acutely bereft of empirical understanding or emanating from single and unrepresentative organizational locations. And with further refinement, along with the generation of even more ambitious aggregate and ethnographic studies elsewhere, perhaps they can be mobilized better to resolve some of the contemporary dilemmas, discussed earlier, in the critical social control literature itself.

As we have seen, even the modest forensic follow-up data presented in this chapter embrace some potentially formidable theoretical discoveries. *First*, they demonstrate an extraordinary breadth and depth of medical and carceral authority when these two institutions are reflexively infused and mutually empowered, which is their typical form of relation when regulating forensic subjects. The frequencies of intervention exhibited in Tables 1, 2 and 4 were indicative of an adrenaline-charged mega-system with an almost manic capacity to process and recycle subjects at blinding tempos within and across multiple control sites. In the six years following their METFORS double labelling, a large proportion of the "mentally disordered offenders" were drawn back in again and again — for an average, per subject, of 4.47 psychiatric interventions, 3.92 carceral contacts, and 8.39 encounters altogether in just six years!

Second, when the frequency of criminal sentences, imprisonments and hospitalizations was juxtaposed against the duration of these interventions, another prominent pattern surfaced. The relations between medico-legal subjects and the institutions that regulate them — both penal and psychiatric — are typically marked by recurrent but relatively brief encounters. Table 2 portrayed a regulatory system in which, over the course of six years, 111 subjects (68.2 percent) received a total of 445 prison sentences, and yet the total average time served per subject was less than ten months. Even more striking, the 91 follow-up inpatients (56.2 percent) endured 321 total admissions, but the total period of confinement for each person averaged just three months. Here we have more dramatic evidence of the "control cycles" (Menzies, 1987, 1989) which became such a dominant attribute of regulatory systems in the 1980s. As fiscal realities have intruded (Lowman and Menzies, 1986; Scull, 1984), and as claims to treatment efficacy and rehabilitative potential diminish, the classic "inclusionary" institutions — prisons and hospitals — have, at least for these 162 people, come to loom in their lives more as frequent or intermittent stop-over centers than as perpetual elements of their control environment. At the same time, as we argued earlier, these developments in no way diminish the power of prisons and psychiatric institutions as the continuing nuclei of legal and medical ordering systems. These remain the substantive core out of which all state coercion, surveillance, treatment and care are ultimately constituted and defined. Deconstructionist extremism, that would privilege the community and dematerialize the state and its control institutions, is grievously at odds with our findings. It is the form, tempo and rhythm of carceral and therapeutic policies, discourses and practices — not their systemic power or their ontological presence — that have changed.

Third, critics of the deinstitutionalization movement, and of the ideas and activities it represents, must assiduously avoid essentializing the "medico-legal system" as some sort of holistic unitary phenomenon. Empirical studies of the kind presented here, which follow people through both penal and therapeutic worlds, can help to underscore the convergences in these formations and their effects — but it is nonetheless imperative that the relative autonomies of law and psychiatry be preserved. While at one level of analysis it may be appropriate to characterize the overall enterprise as an umbrella mega-system, at other planes there remain substantial differences between the subjects, practices and out-comes of carceral and mental health interventions, which are discernible even in the aggregate official data comprising this current study. Penal and mental regulatory strategies were distinct and divergent along a number of dimensions, including the frequency and amplitude of intervention, the relative duration of supervision and confinement, and the relative intensity of surveillance exercised over segregated prisoners and patients (see Tables 3 and 5). Once again, a deconstructionist account that would collapse the boundaries between carceral law and psychiatric medicine, and treat them as equivalent or interchangeable components of a diffused and undifferentiated metasystem, is empirically untenable.

Fourth, the discovered attrition of forensic subjects out of control systems over the years provides a strong antidote to the inevitablism evident in many critical, and particularly "left realist," accounts (see Menzies, 1992). The "wider, stronger and different" nets (Austin and Krisberg, 1981; Cohen, 1985) that sweep the deinstitutionalized society are, it turns out, replete with rents and openings. What was not apparent in our earlier two-year follow-ups of this and another cohort of medico-legal subjects (Menzies, 1989; Webster and Menzies, 1987; Menzies, Webster and Sepejak, 1985a,b) has now been brought plainly into relief. The medico-legal apparatus is not air tight. People do escape. Even the various forms of intermediate community control such as parole, probation and bail show dramatic declines from Years One to Six (see Table 2).[12] Nor are they simply being transinstitutionalized over to psychiatric settings, since these too experience a remarkable temporal attrition. The fact is, quite simply, that "the system of penality" (Garland, 1990) is neither the omnipotent leviathan conjured by some radical control theorists nor the fully dispersed knowledge-power grid of Foucaultian fantasies. It leaks. The power it mobilizes and inscribes on its subjects is incomplete, both hermetically and hermeneutically. Like the path analyses favoured by many structuralist sociologists, when charted in motion it betrays enormous error terms at every stage, each filled up with the movements of prisoners, patients and other subjects in the act of getting out.

And where they go is no mystery. This observation, indeed, leads us squarely to our *fifth* and concluding comment. The background to every table presented in this chapter embodies a "community presence" that is deeply infused in the operations of control institutions and the decisions of their agents. And for the forensic subjects themselves, the community was their dominant locus of living. Of the 11,177 patient/months for which we had reliable data, 8931 months (79.9 percent) were spent on the streets. And at the conclusion of the six-year follow-up, only 16 people were in prison, and nine in hospitals. To this extent, despite the continuing and compelling hold exercised by penal and therapeutic systems, and despite sizable frequencies of apprehended criminal and other misconduct registered during the six-year period, these individuals did ultimately experience a high incidence of deinstitutionalization. As time elapsed, they were more and more likely to find themselves back in more familiar environments, in the community, on the streets.

At this juncture we can only speculate about the long-term meaning of this trend for the METFORS subjects, and for the various systems in which they have been entwined over the years. Given our knowledge about the generally marginal, heavily institutionalized and resourceless backgrounds of these people, and given the dire observations evident in the writings on decarceration, urban poverty and homelessness (Brown, 1985; Curtis, 1986; Dear and Wolch, 1987; Lamb, 1984; Scull, 1984; Torrey, 1988), their (at least temporary) breakout from carceral control centers during the span of their post-METFORS lives is probably no great cause for celebration. For students of social control, the bitter paradox of deinstitutionalization has been that victory means defeat. The overt power structures of the state may be evaded, but in the last instance there is often no escape from the freedom that ensues — a freedom to lead disengaged lives of underemployment, homelessness, drug dependency, petty criminality, mental confusion, alienation and despair in the grim war zones of inner city ghettoes.

To discover how the METFORS subjects, and people like them, turn out in the wider expanse of time would demand energies and resources that are quite beyond our own capacities and those of most researchers. In this chapter, and in some of our other work on the subject (Menzies, 1987, 1989; Menzies, Webster and Sepejak, 1985a,b), we have carried the story forward within the quite restrictive confines of our own possibilities. But there are still gaping holes in the narratives, and much remaining to be told. In this circumstance, the need for integrative, longitudinal, multi-institutional work, both aggregate and ethnographic in quality, has never been more compelling.

If we are ever to comprehend the complex role played by various state institutions in our social world, and to understand how they succeed in exerting regulatory control in our communities and private lives, it is imperative to generate more and better knowledge in critical scholarship about these regulatory practices and their effects. In time, we may come, through research of this kind, better to understand the form and substance of social ordering practices, and their influence on community as well as institutional existence. Such knowledge in turn might eventually inspire some real and realistic initiatives towards establishing more peaceful, humane, non-punitive, de-professionalized and genuinely "communitarian" methods for dealing with those who, for whatever reason, come to violate state and cultural standards of legality, decency and normalcy. Such change, if it ever transpires, will no doubt come far too late to benefit the 162 subjects of this study. But the very tracks that these people (and their analogues elsewhere) leave behind, in the course of their desultory careers through prisons, hospitals, the community, and various other stations of social order, provide important clues to the workings of these systems, and hence to their potential transformation. It is knowledge, therefore, that is well worth pursuing.

Endnotes

1. Funding for this research was supplied by the Social Sciences and Humanities Research Council of Canada, the Canadian Psychiatric Research Foundation, the Solicitor General Canada (from its Unsolicited Research Grants program and from a sustaining grant provided to the University of Toronto Centre of Criminology), the LaMarsh Research Program on Violence and Conflict Resolution, the Ontario Ministry of Health, Simon Fraser University, and the Clarke Institute of Psychiatry. Thanks to Shelley McMain, Shauna Staley and Rosemary Scaglione for co-ordinating data collection at various stages of the project; to Ian Forrester and Vanessa Payne for their bibliographic assistance; and to the many data coders and representatives of mental health, police, justice and correctional agencies who collaborated in the compilation of materials cited in this chapter.

2. Robert Menzies is Associate Professor in the School of Criminology, Simon Fraser University. Christopher D. Webster is Professor and Chair, Department of Psychology, Simon Fraser University.

3. METFORS consists of two facilities: a Brief Assessment Unit (BAU) which delivers one-day psychiatric assessments for the criminal courts, and a 23-bed Inpatient Unit for protracted remands averaging about one month in duration. Since its inauguration, more than 10,000 criminal defendants have undergone forensic examinations in this agency.

4. Initially, assessment teams comprised a presiding psychiatrist along with a forensic psychologist, social worker, psychiatric nurse and correctional officer.

In recent years the operation has been scaled down, with most evaluations being conducted by a psychiatrist, with assistance from the nurse, and the correctional officer where required.

5. These included the offenses of: failure to appear; breach of parole, bail or probation conditions; being unlawfully at large; and escaping lawful custody.

6. None of the 162 persons registered an arrest in the province of Prince Edward Island. Provincial authorities failed to provide correctional data on the three persons who were arrested, respectively, in Manitoba and Nova Scotia. The remaining provincial data base (excluding the home province of Ontario) included 13 subjects arrested in British Columbia, nine in Alberta, six in Quebec, five in Saskatchewan, and one each in New Brunswick and Newfoundland. Although the quality of correctional data varied across provinces, in most cases it was possible to obtain a full profile of institutionalizations and releases, and to gain at least partial information about misconducts incurred during imprisonment.

7. The 162 subjects altogether received 52 additional brief assessments and 55 inpatient evaluations at METFORS during the six-year follow-up. The nine other institutions surveyed (with number of inpatient admissions parenthesized in each instance) were: Queen Street Mental Health Centre (167), Penetanguishene (28), Whitby Psychiatric Hospital (22), Clarke Institute of Psychiatry (18), Lakeshore Psychiatric Hospital (9), St. Thomas Psychiatric Hospital (6), North Bay (2), Kingston Psychiatric (2), and Brockville (1). There were eight admissions to psychiatric wards at general hospitals in the Toronto area. Eighteen additional hospitalizations were identified outside of Ontario.

8. At the conclusion of the 72 months, 105 subjects were at liberty, 24 were on probation or parole, nine were in federal prison, seven were in provincial prison, seven were inpatients in psychiatric hospitals, four were outpatients, and four were dead.

9. For the purposes of data extraction and coding, these were defined as non-violent but disruptive incidents, following as closely as possible the interpretation conveyed by the original author of the relevant document. In the case of conduct occurring in the community and prison, these consisted simply of all transactions leading to criminal charges, hospitalizations or institutional misconducts that were neither violent nor self-injurious. With hospital documents the coding process was somewhat more subjective; coders were instructed to register an incident whenever a patient was implicated in an officially documented overt action that was in some manner seen to be an infraction of formal rules or informal organizational expectations. Operationally, this definition was rather similar to Steinert's (1985) concept of the "problematical situation."

10. Refer to note 5 above.

11.   To generate this statistic, the number of incidents in question, across the entire cohort, was divided by the total number of months spent by all subjects under the relevant institutional or community condition. The quotient was then multiplied by 12 to obtain an annual rate.

12.   As we point out elsewhere, of course, our observations regarding the attrition of subjects and incidents across time may be influenced in unknown ways by changes in the accuracy of the data base over the course of the six-year follow-up period.

References

Almond, R. 1974. *The Healing Community*. New York: Jason A. Aronson.

Ashford, J.B. 1989. "Offence comparisons between mentally disordered and non-mentally disordered inmates," *Canadian Journal of Criminology*. 31: 35-48.

Austin, J. and B. Krisberg. 1981. "Wider, stronger and different nets: The dialectics of criminal justice reform," *Journal of Research in Crime and Delinquency*. 18: 165-196.

Bachrach, L.L. (ed.). 1983. *Deinstitutionalization*. San Francisco: Jossey-Bass.

Bassuk, E.L. and S. Gerson. 1978. "Deinstitutionalization and mental health services," *Scientific American*. 238: 46-53.

Blomberg, T.G. 1987. "Criminal justice reform and social control: Are we becoming a minimum security society?" in Lowman, J., R.J. Menzies and T.S. Palys (eds.), *Transcarceration: Essays in the Sociology of Social Control*. Aldershot: Gower.

Bloom, B. 1973. *Community Mental Health*. Morristown: General Learning Press.

Blumstein, A., J. Cohen, J. Roth and A. Visher (eds.). 1986. *Criminal Careers and Career Criminals*. Washington: National Academy Press.

British Columbia Ministry of Health. 1987. *Mental Health Consultation Report: A Draft Plan to Replace Riverview Hospital*. Victoria: Queen's Printer for British Columbia.

Brown, P. 1985. *The Transfer of Care: Psychiatric Deinstitutionalization and its Aftermath*. London: Routledge.

Burchell, G., C. Gordon and P. Miller (eds.). 1991. *The Foucault Effect: Studies in Governmentality*. Chicago: University of Chicago Press.

Butler, B.T. and R.E. Turner. 1980. "The ethics of pre-arraignment psychiatric examinations: One Canadian viewpoint," *Bulletin of the American Academy of Psychiatry and the Law*. 16: 368-404.

Chan, J. 1992. *Doing Less Time: Penal Reform in Crisis*. Sydney: University of Sydney

Institute of Criminology Monograph Series.

Chunn, D.E. and R.J. Menzies. 1990. "Gender, madness and crime: The reproduction of patriarchal and class relations in a pre-trial psychiatric clinic," *Journal of Human Justice*. 2: 33-54.

Cohen, D. (ed.). 1990. *Journal of Mind and Behavior*. 11 (Special Issue).

Cohen, S. 1985. *Visions of Social Control: Crime, Punishment and Classification*. Cambridge: Polity.

Cohen, S. 1988. *Against Criminology*. New Brunswick NJ: Transaction.

Cohen, S. December, 1991. "Social control and the politics of reconstruction. Strafrecht, soziale Kontrolle, soziale Disziplinierung Konferenz," Bielefeld GER.

Curtis, W.R. 1986. "The deinstitutionalization story," *Public Interest*. 85: 34-49.

Dear, M. and J. Wolch. 1987. *Landscapes of Despair: From Deinstitutionalization to Homelessness*. Princeton: Princeton University Press.

Donzelot, J. 1979. *The Policing of Families*. New York: Pantheon.

Farrington, D.P. and R. Tarling (eds.). 1985. *Prediction in Criminology*. Albany: State University of New York Press.

Foucault, M. 1977. *Discipline and Punish: The Birth of the Prison*. New York: Pantheon.

Foucault, M. 1978. "About the concept of the 'dangerous individual' in 19th-century legal psychiatry," *International Journal of Law and Psychiatry*. 1: 1-18.

Friedenberg, E.Z. 1976. *The Disposal of Liberty and Other Industrial Wastes*. New York: Doubleday.

Garland, D. 1985. *Punishment and Welfare: A History of Penal Strategies*. Aldershot: Gower.

Garland, D. 1990. *Punishment and Modern Society: A Study in Social Theory*. Chicago: University of Chicago Press.

Garland, D. and P. Young. 1983. *The Power to Punish: Contemporary Penality and Social Analysis*. London: Heinemann.

Gibbens, T.C.N., K.L. Soothill and P.J. Pope. 1977. *Medical Remands in the Criminal Court*. Oxford: Oxford University Press.

Hall, H.V. 1987. *Violence Prediction: Guidelines for the Forensic Practitioner*. Springfield: Charles C. Thomas.

Hodgins, S. (ed.). 1993. *Crime and Mental Illness*. Newbury Park: Sage.

Kiesler, C.A. 1987. *Mental Hospitalization: Myths and Facts About a National Crisis*. Newbury Park: Sage.

King, R.D., N.V. Raynes and J. Tizard. 1971. *Patterns of Residential Care*. London: Routledge.

Lamb, H.R. (ed.). 1984. *The Homeless Mentally Ill*. Washington DC: American Psychiatric Association.

Lerman, P. 1982. *Deinstitutionalization and the Welfare State*. New Brunswick NJ: Rutgers University Press.

Lewis, D.A. 1991. *Worlds of the Mentally Ill: How Deinstitutionalization Works in the City*. Carbondale: University of Illinois Press.

Lowman, J. and R.J. Menzies. 1986. "Out of the fiscal shadow: Carceral trends in Canada and the United States," *Crime and Social Justice*. 26: 95-115.

Lowman, J., R.J. Menzies and T.S. Palys (eds.). 1987. *Transcarceration: Essays in the Sociology of Social Control*. Aldershot: Gower.

Mandel, M. 1991. "The great repression: Criminal punishment in the nineteen-eighties," in Samuelson, L. and B. Schissel (eds.), *Criminal Justice: Sentencing Issues and Reforms*. Halifax: Garamond.

Mathiesen, T. 1980. "The future of control systems — The case of Norway," *International Journal of the Sociology of Law*. 8: 149-164.

McMain, S., C.D. Webster and R.J. Menzies. 1989. "The post-assessment careers of mentally disordered offenders," *International Journal of Law and Psychiatry*. 12: 189-201.

Melossi, D. 1990. *The State of Social Control: A Sociological Study of Concepts of State and Social Control in the Making of Democracy*. New York: St. Martin's Press.

Menzies, R.J. 1987. "Cycles of control: The transcarceral careers of forensic patients." *International Journal of Law and Psychiatry*. 10: 233-249.

Menzies, R.J. 1989. *Survival of the Sanest: Order and Disorder in a Pre-Trial Psychiatric Clinic*. Toronto: University of Toronto Press.

Menzies, R.J. 1992. "Beyond realist criminology," in Lowman, J. and B.D. MacLean (eds.), *Realist Criminology: Crime Control and Policing in the 1990s*. Toronto: University of Toronto Press.

Menzies, R.J., C.D. Webster, S. McMain, S. Staley and R. Scaglione. In press. "The dimensions of dangerousness revisited: Assessing forensic predictions about violence," *Law and Human Behavior*.

Menzies, R.J., C.D. Webster and D.S. Sepejak. 1985a. "Hitting the forensic sound barrier: Predictions of dangerousness in a pre-trial psychiatric clinic," in C.D. Webster, M.H. Ben-Aron and S.J. Hucker (eds.), *Dangerousness: Probability and Prediction, Psychiatry and Public Policy*. New York: Cambridge University Press.

Menzies, R.J., C.D. Webster and D.S. Sepejak. 1985b. "The dimensions of dangerousness: Evaluating the accuracy of psychometric predictions of violence among forensic patients," *Law and Human Behavior*. 9: 35-56.

Miller, P. and N. Rose (eds.). 1986. *The Power of Psychiatry*. Cambridge: Polity.

Moore, M., S.R. Estrich, D. McGillis and W. Spelman. 1984. *Dangerous Offenders: The Elusive Target of Justice*. Cambridge: Harvard University Press.

Nuffield, J. 1982. *Parole Decision Making in Canada: Research Toward Decision Guidelines*. Ottawa: Supply and Services Canada.

Olsen, R. (ed.). 1979. *Alternative Patterns of Residential Care for the Discharged Psychiatric Patient*. Birmingham: BASW.

Pasamanick, B., F.R. Scarpitti and S. Dinitz. 1967. *Schizophrenics in the Community*. New York: Appleton.

Pearson, G. 1975. *The Deviant Imagination: Psychiatry, Social Work and Social Change*. London: Macmillan.

Ralph, D. 1985. *Work and Madness*. Montreal: Black Rose.

Rose, N. 1990. *Governing the Soul: The Shaping of the Private Self*. London: Routledge.

Scull, A. 1984. *Decarceration. Community Treatment and the Deviant — A Radical View*. 2nd ed. New Brunswick NJ: Rutgers University Press.

Steadman, H.J. 1979. *Beating a Rap? Defendants Found Incompetent to Stand Trial*. Chicago: University of Chicago Press.

Steadman, H.J., P.C. Robbins and J. Monahan. 1991. "Predicting community violence among the mentally ill." 50th Annual Conference of the American Society of Criminology, San Francisco.

Steinert, H. 1985. "The amazing new left law and order campaign," *Contemporary Crises*. 9: 21-35.

Talbott, J. (ed.). 1984. *The Chronic Mental Patient Five Years Later*. New York: Grune and Stratton.

Teplin, L.A. (ed.). 1984. *Mental Health and Criminal Justice*. Newbury Park: Sage.

Thornberry, T.P. and J.E. Jacoby. 1979. *The Criminally Insane: A Community Follow-up of Mentally Ill Offenders*. Chicago: University of Chicago Press.

Toch, H. and K. Adams. 1989. *The Disturbed Violent Offender*. New Haven: Yale University Press.

Torrey, E.F. 1988. *Nowhere to Go: The Tragic Odyssey of the Homeless Mentally Ill*. New York: Harper and Row.

Webster, C.D. and R.J. Menzies. 1987. "The clinical prediction of dangerousness," in Weisstub, D.N. (ed.), *Law and Mental Health: International Perspectives*. Vol. 3. New York: Pergamon.

Webster, C.D. and R.J. Menzies. 1989. "Violence and mental illness," in Wolfgang, M.E. and N.A. Weiner (eds.), *Pathways to Criminal Violence*. Newbury Park: Sage.

Webster, C.D., R.J. Menzies and M.A. Jackson. 1982. *Clinical Assessment Before Trial: Legal Issues and Mental Disorder*. Toronto: Butterworths.

Wing, J.K. and R. Olsen. 1979. *Community Care For the Mentally Disabled*. Oxford: Oxford University Press.

Confronting Individual and Structural Barriers to Employment: The Employment and Skills Enhancement (EASE) Program for Prisoners

Denis C. Bracken and Russell J. Loewen

It is tempting to link unemployment to crime. The idea is a compelling one that if someone is not involved in some kind of "productive" labour, they are up to no good. In his historical study of unemployment, Garraty suggests that the unemployed have always been considered as potential criminals, and that proposed methods to ameliorate the problem of unemployment have been around almost as long as the problem itself.

> Unemployed persons have been treated as criminals who must be isolated from society or driven to hard labour, and as sinners to be regenerated by exhortation and prayer.... Nearly every scheme for both improving their lot and sustaining them in their misery that is currently in vogue, along with many no longer considered workable, was known and debated at least as far back as the 16th century. What actually has been done for the unemployed and about unemployment has depended upon the interaction of moral and religious attitudes, the sense of what is economically possible, the locus of political power in society, and the extent to which those who possess the power are aware of how unemployment affects both its victims and their own interests (Garraty, 1978 p. 9).

Conversely, one might say that if someone is, in fact, involved in productive labour, then chances are that he or she is quite unlikely to be participating in criminal activity. Indeed, the image of the criminal as someone who is not "working," in the traditional sense of the term, is an enduring one (Chambliss, 1964). Some would argue that the development of the prison in the 19th century was strongly influenced by the need to reinforce the routines of industrial discipline necessary for emerging industrial societies (cf. Rothman, 1990). As such, finding employment for people "at risk" of being involved in crime (particularly if they have been convicted of criminal activity before) has been a major part of efforts to keep people out of jail for over 100 years. As Rothman pointed out in the American context of the last century:

> The commitment to a daily routine of hard and constant labour...pointed to the close correspondence between the ideas on the causes of crime and the structure of the penitentiary. Idleness was part symptom and part cause of deviant behaviour. Those unwilling to work were prone to commit all types of offences; idleness gave time for the corrupted to encourage and instruct one another in a life of crime (Rothman, 1990, p. 103).

What we attempt to do in this chapter is to examine a program in place in two provincial prisons in Manitoba which tries to address directly a prisoner's chances of finding employment upon release in the context of looking at both structural barriers to employment, and personal ones. We consider this first within a context of some summary research on the relationship between unemployment and imprisonment, particularly with respect to the notion of recidivism, i.e., the likelihood of prisoners committing new offences following release from custody.

The idea that crime is likely to increase as unemployment increases is one initially considered by Rusche (1933). The needs of a labour market are thought to have an impact on the number of crimes committed, and, concomitantly, the number of people receiving convictions and prison sentences. Direct correlations between crime rates and unemployment rates, however, have not been well established.

In a review of the literature, Box and Hale (1986) found that "The best available evidence...does not provide unambiguous support for the hypothesis that unemployment causes more crime..." (p. 77). A subsequent review of the literature came to essentially the same conclusion (Crow, Richardson, Riddington and Simon, 1989, p. 11). What Box and Hale do suggest, however is the importance of the enduring perception:

> ...it is clear that many people *believe* that unemployment causes crime and this belief has real consequences, particularly when it affects decisions taken by state officials processing suspected and convicted persons (p. 72).

Indeed, they suggest that the evidence supporting this conclusion is easier to find. Reviewing several studies on factors employed in sentencing, they find that the employment status of the offender at time of sentencing will influence the likelihood of a custodial sentence. They believe that although research suggests no simple relationship between incarceration and crime rates, it "fails to dissuade the judiciary from using their commonsense notions of crime/causation to guide them in sentencing unemployed males" (p. 83).

In an extensive study of the relationship between unemployment and criminal offenders in the United Kingdom, Crow, Richardson, Riddington and Simon (1989) point out that what happens to an offender at various stages of the criminal justice process (e.g., bail, sentencing, parole) is quite likely to be influenced by whether or not the offender is employed or has a chance at employment. When one considers, as they do, that offenders by and large come from those segments of the population most often unemployed, the issue becomes an important one. They found that employment-related approaches to "rehabilitation" often neglect the situation of the labour market, and instead focussed on the perceived needs of the offender population for some type of training with the dual hope of getting these people employed and reducing the likelihood of their return to prison. There was very much a linkage, in the minds of these authors, between the presumed individual shortcomings of offenders and employment-related rehabilitation programs.

> ...people convicted of offences tend to come from sectors of the population who are most at risk of unemployment or are disadvantaged in the labour market because they have few qualifications. One consequence of this was that in the past schemes for unemployed offenders were seen as rehabilitative, in the sense that their aim was to compensate for these deficiencies and to re-integrate offenders into what was taken to be normal working life.... Another feature of such schemes was that the emphasis was very much on the individual offender, rather than the needs of the labour market (p. 76).

Someone interested in working with prisoners on the issue of employment and unemployment, then, is faced with a daunting task. The "normal" correctional approach, as Crow et al. outline above, has been to assume that unemployment and its attendant problems are the result of deficiencies within the prisoner population, and the segment of society from whence they have come. As they point out, the larger issues related to social and economic causes of unemployment have largely been considered marginal to criminal justice. Bringing these issues back from the margins, so to speak, as part of a program of working with prisoners on the issue of unemployment, might be one approach. As such, it might involve a lessening of the goal of most employment programs for offenders of simply finding a job and not returning to prison. It might include an awareness of the personal impact of one's position in society on how one communicates, deals with stress, and looks for some meaningful and constructive relationship to one's community, be it through employment or something else. The endeavour here is to attempt to utilise an approach similar to the "social analysis of penality" described by Garland and Young (1983). More specifically, working with prisoners on problems like employment can possibly allow for bridging the gap between those "...institutions which 'deal with offenders'...[and] those which can intervene to transform the offenders' social conditions" (Garland and Young, 1983, p. 34). The EASE Program of the John Howard Society of Manitoba has attempted to address a small segment of these much larger concerns.

During the summer of 1990, the John Howard Society of Manitoba evaluated the services it provided to the Headingly Correctional Institution, the largest provincial prison in Manitoba. A number of institutional staff and agency service providers were interviewed to determine, in part, those areas of greatest need experienced by prisoners. The Employment and Skills Enhancement (EASE) pilot project was developed and implemented as a result of the 1990 agency evaluation. It was intended for prisoners who, for a variety of reasons, had not experienced successful employment in the community. It has since been offered in a modified form at the Brandon Correctional Institution by the John Howard Society in Brandon.

Adult institutional corrections in Manitoba are centred largely around the two main provincial prisons, Headingly (just west of Winnipeg) and Brandon, although there are smaller institutions in Dauphin, The Pas and Milner Ridge near the town of Beausejour. Headingly and Brandon are the most secure institutions with the largest percentage of the prison population. To put this in perspective, it is worth noting the following information about adult corrections in the province. Total sentenced admissions in 1990 to Manitoba correctional facilities were 5148, of whom 49 percent were native. The median sentence was 90 days and the median age on admission was 28 (Canadian Centre for Justice Statistics, 1991). Institutional programming is limited.

The EASE program begins with the assumption that obtaining and maintaining employment is a major factor in preventing prisoners from receiving a custodial sentence should they be rearrested after release. Underlying assumptions about individual employment skills include the need for self-esteem, motivation, and courage building. EASE attempts to address issues such as street support, conflict resolution (anger and stress management), literacy, résumé writing, and self-presentation—all barriers to individuals finding and maintaining employment. In addition, EASE also focusses on the structural barriers to employment that prisoners often face on release.

EASE also serves as a way to enhance other employment-related programs on the outside, by providing an entry level program on employment issues. The John Howard Society of Manitoba operates a six week pre-employment course entitled the Social Skills Orientation Course (SSOC). Part of one EASE session is devoted to a description of SSOC. Those individuals who would like to attend the community-based program may discuss that with the SSOC presenter. The John Howard Society of Manitoba also provides a community-based Employment Assistance Service (EAS). The EASE worker discusses this service and thus serves as a community link. Additionally, C.E.I.C. presenters discuss U.I. benefits and eligibility. This linkage to the community through outside agency participation provides an avenue for breaking down of stereotypes, awareness raising, and the sense that the community may have something to offer in terms of the welfare of people in prison.

The EASE program is open to virtually any prisoner at Headingly or Brandon, although it is primarily geared for individuals who have some job skills but have failed to maintain regular employment while on the street. Prisoners are informed about the program through their Unit Counsellors, members of the inmate counsel, or John Howard Society workers at the institution, and given the opportunity to participate if they wish. Selection is undertaken by the EASE co-facilitators; participants are asked to indicate a willingness to participate actively in the program, attend the nine sessions over three weeks, and to be prepared to address personal barriers to employment. A total of 50 prisoners have participated in the program over three sessions at the two institutions. The sessions are described in the chart given below:

1: Introduction	Program co-facilitators	Intro. to program. Discussions on anger and implications for work setting
2: Stress	Program co-facilitators	Stress in the workplace and problem solving
3: Communication and conflict	Program co-facilitators	Discussion on communication and conflict
4: Community programs	Program co-facilitators and two staff from local literacy program	How to make use of community programming, discussion on available literacy resources in the community

5: Logistical aspects of employment	JHS worker	Résumés, SINs, birth certificates, bonding, application forms, etc.
6: Conflict resolution I	Worker from the Community Dispute Centre	Role plays dealing with non-violent conflict resolution in the workplace
7: Government-sponsored programs	Canada Employment Immigration Commission worker	Information on unemployment insurance, employment opportunities and programs through CEIC
8: Job searching and pre-employment programs	JHS Employment counsellor, and two members of the business community; SSOC staff member	Discussion on looking for a job, and comments from business people on whom they hire, and why
9: Closing session	Program co-facilitators	Review of the previous sessions, feedback and completion of feedback instruments, diplomas and applause

The program has been under way for a short time, and the total number of participants has been small. Resources have not been available to date for any extensive review of the program, nor for post-release follow-up with participants. However, information has been gathered at the end of each session to assist in planning for additional sessions. This has involved asking participants to complete questionnaires related to the content of the sessions, using scales to focus answers across a range from most helpful/beneficial to least. As well, other questions provide for more open-ended answers concerning the sessions. This data, rudimentary though it is, would tend to suggest that virtually all participants found the sessions to be useful, particularly those in stress management and the practical aspects of employment.

Much of current correctional programming involves keeping people within the walls of the prison until either an early departure has been earned, or the expiry of the penal sentence. Freedom in the sense of release from prison means a "return" to the society from which the individual came, but likely that individual will be no better equipped to deal with the challenges that one is confronted with in society each day. One challenge of the released prisoner probably will be the prospect of participating in some productive manner in an economy which increasingly requires more than the ex-prisoner has to offer in the way of skills. It is beyond the scope of the EASE program to overcome this challenge. However, it is important that prisoners have some ability both to cope with these challenges, and to understand why these challenges are arising. To do this, the EASE Program attempts to promote values like self-determination and the view that freedom is strongly dependent on economic and political equality. The participants of the EASE program are encouraged to consider personal and structural barriers to employment. Blame is not placed on the individual who, perhaps for reasons related to race or socio-economic class, finds himself unable to secure employment. Participants are encouraged to consider those structures in society (political, racial, economic) which prevent equal access to employment. This perspective elevates the participant to take responsibility for those things which he can change while locating responsibility for structural barriers to employment where they belong—not within him.

The four goals of incarceration—deterrence, punishment, protection and rehabilitation—are inherently value laden. Persons who enter prison become keenly aware that if they are to survive the system, they must construct their lives according to these system goals. These sentencing goals are ignored within the EASE program because they are in fact incompatible with the program's overall objectives. The goals of sentencing imply that autonomy is taken away from the prisoner for the prisoner's own good, the system then dictating what is right or wrong for the prisoner. This is not a climate under which anyone is willing to address barriers to employment.

A fundamental assumption of the program is that people have the capacity to change certain aspects about themselves, including those which may seem incompatible with employers or co-workers. For many, this change does not occur due to a lack of self-confidence. The EASE program attempts to value individuals for who they are, and without blaming people for where they have come from. The hoped-for result is that there is less reason for participants to remain defensive about personal barriers to employment. As well, there is then an attempt to differentiate between structural and personal barriers. Skills in conflict resolution and problem solving may prove helpful in resolving certain interpersonal relationship issues, but such skills are not recognized as necessarily the solution to more structural issues like inadequate education. For these, the group discussions in EASE encourage consideration of appropriate social action, not self-blame. With this, a greater sense of control over one's life may be developed among participants.

Group discussions within the EASE program also encourage participants to recount their barriers to employment, not necessarily for the purpose only of changing those barriers, but primarily for the value of identifying a common experience. When an individual discusses how anger or conflict has impeded successful employment, he has a great deal in common with other group members. Further group discussion encourages the discovery of the causes of that anger and also how it can be expressed as a positive emotional expression. The participant's anger is affirmed with the group as a natural and healthy emotion which he can take responsibility for and learn to focus.

The goal of EASE is not to rehabilitate. The temptation, however, is to make it fit within current correctional programming in order to fulfil the demands of that system. That elusive carrot of early release dangles ever so closely when one can say that the faults of the individual (i.e., his unemployment) are responsible for imprisonment in the first place. Prison shelters prisoners from life's realities, teaches them to cope by manipulation and coercion, and fails to foster a sense of control over one's own life. Prison employment programs have traditionally been self-serving. Prison "crews" pick garbage, cut trees, clean parks, and harvest vegetables. Inherent is a hierarchical model which maintains a philosophy of divisions between us and them.

Will programs for prisoners such as the Employment and Skills Enhancement meet the real needs of people taking the programs? Whether prison programming can achieve any short term goals without compromising a longer term goal of radically reducing prison populations is questionable. However, there is some argument for suggesting that the traditional measures of success, i.e., reducing recidivism among released prisoner populations, ought not to be the only measure of assessing what a program does. If one goal of EASE is to foster an awareness of the personal impact of one's position in society on how one communicates, deals with stress, and looks for some meaningful and constructive relationship to one's community, then how well this is accomplished might provide a better indication of a program's value or worth. Stan Cohen (1985) suggests that there are issues to be considered when assessing the value of some initiatives other than simply the question of does a given policy or program reduce crime. He states his view that

> ...moral values which are cherished as ends in themselves, should not
> be relabelled as 'means' for the instrumental enterprise of crime
> control. Doing so would only be to devalue these values, and to lead
> to their abandonment if the official purposes of the system are then
> not achieved (p. 265).

Our belief is that the EASE program addresses the issue that the imprisonment of people is inextricably linked to politics and economics. The strength of the program lies in the participation of the prisoners, where people are not judged or blamed on the basis of their individual race or economic status, but where they have the opportunity to find strength as a group of people with common barriers to employment.

References

Box, S. and Hale, C. 1986. "Unemployment, crime and imprisonment, and the enduring problem of prison overcrowding," in Maths, R. and Young, J., *Confronting Crime*. London: Sage.

Canadian Centre for Justice Statistics. 1991. *Adult Correctional Services in Canada 1990-1991*. Ottawa: Statistics Canada.

Chambliss, W. 1964. "A Sociological Analysis of the Law of Vagrancy," reprinted in Carson, W.G. and Wiles, P., 1971, *The Sociology of Crime and Delinquency in Britain*. Volume 1. London: Martin Robertson.

Cohen, S. 1985. *Visions of Social Control*. Cambridge: Polity Press.

Crow, I., Richardson, P., Riddington, C., and Simon, F. 1989. *Unemployment, Crime and Offenders*. London: Routledge.

Garland, D. and Young, P. 1983. "Towards a Social Analysis of Penality," in Garland, D. and Young, P., *The Power to Punish*. London: Heineman.

Garraty, J. 1978. *Unemployment in History: Economic Thought and Public Policy*. New York: Harper and Row.

Rothman, D. 1990. *The Discovery of the Asylum*. 2nd edition. Boston: Little, Brown and Co.

Rusche, G. 1933. "Labour Markets and Criminal Sanction," reprinted in *Crime and Social Justice*. 10, Fall-Winter 1978.

Thomas, J. and Boehlefeld, S. 1991. "Rethinking Abolitionism: 'What do we do with Henry?' Review of de Haan, *The Politics of Redress*." *Social Justice*. 18, 3, Fall p. 238-251.

A Model for Community Corrections:
The Vitanova Recovery Program

Franca Damiani Carella

The recovery program conducted at The Vitanova Foundation's Drug Rehabilitation Centre in Woodbridge, Ontario, is a day and evening outpatient program for substance users, their families, and/or significant others. The client population consists of men and women with alcohol and drug-related problems, including a large proportion of court-ordered referrals. This article explains the details of this holistic, individualized and culturally-sensitive program.

Introduction

The Vitanova Recovery Program, while rooted in a solid understanding of the cultural values and norms of the Italian Canadian community, now serves a diversity of cultural communities in the greater Toronto area. Indeed, its components have wide cross-cultural application. In this regard, the involvement of the substance user's family is key, as is necessary in the context of these communities. For example, many of the parents of substance users have difficulty understanding the issues involved in such use because of cultural and/or linguistic barriers. These parents often came from societies whose social norms were, and may still be, considerably different from those that their children must now face. These barriers cause substantial difficulty in addressing a substance use problem, making community education a priority. However, education is only a first step in what must be a broadly-based continuum of care designed to serve cross-cultural youth and young adults while enhancing the community identity which is crucial to their recovery. This is particularly noteworthy since crises in cultural identity frequently precede the onset of substance use.

Background

The history of The Vitanova Foundation begins with *VitaSana* (Latin for "healthy life"), an Italian-language health magazine (subsequently printed in both Italian and English) first published in Toronto in 1984. The creation of Dr. Renzo Carbone and myself, *VitaSana* was designed to inform non-English-speaking Italian Canadians about individual, family and community health issues.

Substance use in the Italian Canadian community was a recurring topic of *VitaSana* articles from its inception. As a result of these articles, *VitaSana* received an increasing number of enquiries from readers concerned about how the issue could best be handled by parents and families. The response to these enquiries was the VitaSana Family Network, a set of informal support groups designed to help parents cope with the impact of substance use in their own families.

It was soon clear that this informal structure needed expansion in order to provide the range of interventions required to confront the problem. Therefore, in 1987, I invited a number of leading Italian Canadian professionals and business persons in the Greater Toronto Area to form a board of directors. We applied for a non-profit corporate charter, which was granted in 1988. The following year, registered charitable status was awarded by Revenue Canada. In June of 1991, The Vitanova Foundation opened the Drug Rehabilitation Centre, a 9,000-square-foot facility near the intersection of Highways 7 and 400 in Woodbridge, Ontario.

Program Description

The Vitanova Recovery Program provides, in English and six other languages, a culturally-sensitive program offering a continuum of services (see Figure 1) including primary prevention, secondary intervention, orientation, in-depth assessment, tertiary intervention (treatment and rehabilitation) and continuing care to Italian Canadians and other ethnocultural groups resident in York Region and other parts of the Greater Toronto Area. Clients are referred by medical doctors, lawyers, clergy, teachers, guidance counsellors, corrections officers, the courts and others. As well, Vitanova offers a number of services to clients' families and/or significant others.

Program Philosophy

The Vitanova Recovery Program promotes re-entry and reintegration into mainstream society on the latter's terms. High regard is given to honesty as well as personal commitment to regaining health, restoring self-confidence, developing a sense of dignity and purpose, and obtaining full-time employment. Education and training for future employment are emphasized.

Reintegration on society's terms is illustrated by the case of a client who stole several thousand dollars worth of video equipment from our centre. The client was confronted and offered a choice: to be charged with the theft or to enter a high-security residential treatment program, followed by re-entry into our program for the balance of his original two-year commitment, and full restitution. The client chose the latter. We demand, and most usually get from our clients, such commitment, knowing that in regard to substance dependency nothing can be achieved without a genuine commitment to accepting society's rules and a readiness to pay the consequences of failing to do so. We view our task as helping those in need to learn a new way of living, and having learned it, to enjoy the possibilities offered by a new life or *vita nova*.

Michael: A Case Study

Michael was the first person with a history of drug-related criminal charges to enter the Vitanova Recovery Program. Aged 17 at the time, he had been using drugs for more than four years. To finance his drug use, he engaged in petty theft and drug trafficking for most of those years. His first victims were members of his own family, followed by family friends and his own acquaintances, and only later strangers. Discovery of his thefts by family members was followed by a period of bargaining and bribery, including offers of expensive gifts, in exchange for promises to cease his use of drugs. One offer was of a new automobile, which Michael accepted immediately, knowing it would extend the range of his search for goods worth stealing and conveying to dealers in such commodities. With the windows of the automobile tinted, he could carry on his business with discretion, and (except for the coldest months) he need not return home to sleep. Lastly, it gave him a ready source of saleable parts if he were desperate for funds.

The slow realization by Michael's parents that the gift of an automobile was a mistake was made obvious by his arrest and conviction on two charges of "theft under." At the request of his parents who were referred to Vitanova by their parish priest, I appeared on Michael's behalf before the court, resulting in his sentence being suspended on condition he enter the Vitanova Recovery Program for a period of two years, with the additional stipulation that his failure to do so would result in incarceration for the same length of time. Michael agreed, and is now completing his third year of involvement with Vitanova, intending to pursue a career as a certified addiction counsellor.

Michael's parents remain involved as well, as active Vitanova volunteers. They have spent a similar period of time availing themselves of the family education and counselling, individual and group therapy and support group components of the Vitanova Recovery Program, in order to understand better the dynamics of enabling and co-dependent behaviors.

Treatment

The treatment process is preceded by: orientation to the goals and philosophy of the Vitanova Recovery Program; in depth assessment which includes an initial interview with an intake counsellor; a second interview during which a detailed, computer-based questionnaire is completed; medical, psychological and/or psychiatric assessments, where indicated; and contact with the referring individual or agency. Clients are advised that Vitanova does not accept those unable or unwilling to commit freely to attending the program (day, day or evening, or evening) for a period of up to two years. They are then assigned to one of our multidisciplinary team of counsellors (which includes a psychologist, a social worker and a consulting psychiatrist).

Married clients and those still living with their families of origin are encouraged to participate in marriage and family counselling sessions with their significant others, for their mutual benefit and to further client self-understanding. In this regard, it is important to note that in a large proportion of these cases, some manner of physical, emotional, mental or sexual abuse is reported.

Treatment for heroin users begins with off-site detoxification, for a period of from seven to ten days. A frequent locale is the detoxification centre connected with a nearby hospital. This facility is preferred by heroin users, who, in general, fear the physical side-effects of detoxification and appreciate the proximity of emergency medical services.

Treatment for cocaine users centres on careful observation and in-house detoxification generally within the first three months following the start of program participation. This is dictated by the fact that the urge to use varies from individual to individual: for some the habit is daily, for others weekly, twice-monthly, or even less frequently. But almost all cocaine users relapse once in a three-month period. Hence the length of this period of intense observation, sufficient time to reveal the pattern of recurring urge for use and to provide the necessary intervention.

Treatment for crack-cocaine users is quite similar to that for cocaine users, with an added feature: as some measure of brain trauma is associated with crack-cocaine use, assessment by our consulting psychiatrist is nearly always recommended. One problem we frequently encounter is the apparent inability of recovering crack-cocaine users to retain new information for any length of time, a serious matter when attempting to teach new behaviors and the reasons for adopting them. Thus, we have begun to explore opportunities for the cognitive retraining of these clients, along much the same lines as the retraining programs developed for those who have suffered moderate brain damage as a result of automobile accidents.

Rehabilitation and Re-entry

Rehabilitation is geared to the learning of a range of life, social, and job skills facilitating re-entry into the mainstream of society.

Thus, a typical day at the Drug Rehabilitation Centre begins with vigorous physical exercise in our gymnasium: calisthenics, plus use of a step machine, a treadmill, a stationary bicycle or a rowing machine. Beside the obvious aerobic benefits, regular exercise teaches physical self-discipline and helps develop a positive body image.

Exercise is followed by group counselling, and individual counselling when necessary. The program continues with a work therapy session in a small on-site ceramic factory, affording clients the opportunity to express a measure of their own creativity, to experience the discipline of regular work, and to enjoy the sense of achievement based on tangible results. The sale of ceramic products is intended to support future job training initiatives, such as the outfitting of a computer classroom.

Lunch is prepared by and served to clients under the direction of a staff member or volunteer. Besides learning how to cook and serve nutritious meals, it is an opportunity to learn how to play host to others who, in turn, learn how to dine in a mannerly fashion and to clean up after themselves.

Mealtime is followed by a period of meditation, then reflection and discussion based on the twelve-step model and a second group therapy session.

The final half-hour of the day is taken up with maintenance tasks: vacuuming carpets, washing floors, dusting furniture, putting everything back in its place, work that is valuable in its own right, as well as teaching cleanliness, order and respect for one's surroundings.

Short-term Shelter

In some cases, when clients live with parents who refuse to participate in our program, and if those parents are especially enabling, successful recovery can only be achieved by the client removing himself from that environment. Hence, our "house," not a residential treatment center so much as a "home" for those whose living situation hinders recovery. Two such situations recently encountered involved young men, one of whose father both uses and deals in drugs, and another whose sibling refuses to give up his substance use and accept the help offered to him.

Re-entry Outcomes

Several clients are now completing the requirements for diplomas as certified addiction counsellors. Others have started their own businesses after apprenticing for a time with employers who were willing to hire them on our recommendation. In such instances, we make certain both employer and employee know we are available to help when and if problems arise.

Future Directions

Current concerns focus on maximizing time spent by staff in direct contact with clients and, to permit that, developing a means of monitoring every aspect of program participation at a minimum investment of staff time. Thus, we are developing a computer-based program which, besides allowing individual case logging, will record client participation in program components, register discrete assessments of participation levels in such components, flag lapses participation, and maintain client contact following program completion. Ultimately, we view computer-based montioring as an opportunity to apply cost-benefit analysis to a wide range of community-based services, while developing at the same time a data base for future research purposes.

Summary

Given Canada's multicultural context, effective treatment and rehabilitation of individuals with alcohol and drug-related problems requires a range of orientation, assessment, treatment, rehabilitation and follow-up services which are sensitive to the individual's, and his or her family's, cultural context. The Vitanova Recovery Program offers just such a continuum of culturally-sensitive, client-driven services. In keeping with The Vitanova Foundation's status as a charitable organization, these services are offered without charge.

A Legacy of Colonialism and Racism: Aboriginal Injustice in Australia

Robert Doyle and Linda Freedman

> In short, an Aborigine is much more likely than a white in Australia
> to be in one or more of the following states: sick, unemployed,
> uneducated, poor, imprisoned or dead (Haupt, 1987: 6).

Introduction: Year of Indigenous Peoples

Another year for another United Nations target group! After witnessing in recent years the status and problems of such groups as the homeless and children, the focus in 1993 is on the plight of indigenous peoples. However, despite being highlighted by the U.N., target groups from other years do not seem to have benefited enormously in a world in which extreme ethnic and tribal violence is now a core characteristic. Will a 'year of indigenous people' mean anything positive for indigenous people in countries such as Australia (with a total population of 17 million) and Canada (of 27 million)? Will they benefit in any way from being elevated to a focus of international attention and/or curiosity? Only time will tell, but if one were a 'punter' in Australia, the odds would not favour making a 'large punt' that the situation of disadvantage for Aborigines in their own country will be substantially improved in 1993. We intend to describe the racism and appalling treatment of Aboriginal people in Australia since the invasion of European colonialists (which show some similarities with other indigenous people), and analyze the current status of Aboriginal disadvantage. We will outline some major initiatives being taken by governments (principally the Commonwealth Government) in this country, as well as some thrusts by Aboriginal people to promote a greater sense of 'community control.' Lastly, we will look, as a case example, at the treatment of Aboriginal people in the justice system which currently does little to serve them.

Firstly, however, it is important to discuss what is meant by 'indigenous people.' The task is not as easy as it appears since indigenous people have been reluctant to define themselves because of fears of colonialist governments which have most often sought to colonize and oppress them. The World Council of Indigenous Peoples (WCIP) use a political definition, for official purposes, which is itself insufficient because it does not address special cultural characteristics or distinguish indigenous peoples from any national ethnic minorities who are 'native' to a country.

> Indigenous people shall be people living in countries which have populations composed of different ethnic or racial groups who are descendants of the earlier populations which survive in the area, and who do not, as a group, control the national government of the countries within which they live (WCIP Information leaflet).

Reid and Lupton (1991) describe "fourth world" or indigenous populations as characterized by their experience of being colonized, or of being a minority in relation to a dominant, encompassing state. Many of these peoples have been forced to assimilate (or dominant governments have tried to assimilate them and failed), in the process losing their land and their economic base, and therefore their autonomy. The Independent Commission on International Humanitarian Issues (1987: 7) uses a working definition of indigenous people as

> ...composed of the existing descendants of the peoples who inhabited the present territory of a country wholly or partially at the time when persons of a different culture or ethnic origin arrived there from other parts of the world, overcame them and, by conquest, settlement or other means, reduced them to a non-dominant or colonial situation; who today live more in conformity with their particular social, economic and cultural customs and traditions than with the institutions of the country of which they now form a part, under a state structure which incorporates mainly the national, social and cultural characteristics of other segments of the population which are dominant.

It is also noteworthy that, according to the WCIP definition, if indigenous peoples were to be recognized as independent, sovereign nations, controlling their own national governments, that they would cease to be indigenous people (Bodley: 153). In this article, we accept the limitations of the WCIP definition for Australia and acknowledge the movement of indigenous people in this

country, the Aborigines and Torres Strait Islanders, to assume greater community control and to seek recognition of their sovereignty and self-governing
capacity within the Australian nation state. One of the ongoing difficulties,
particularly for European colonialists and their descendants, is what to call
indigenous people. In Canada, "Indian" as noted below is a legal term, and many
indigenous people prefer the term "First Nations" which separate them from
those who took and occupy their land. In Australia, "Aborigine" is not a
universally accepted term. In different parts of the continent, specific terms are
employed. For example, Fesl (1987) notes:

> 'Koori' is the term by which the indigenous people of Australia living
> along the eastern and southern parts of Australia are known to each
> other. In 1988 it was decided that the term 'aborigine', an identityless
> label applied by the colonists, would be discarded and the term 'Koori'
> used publicly. The term means 'our people'; it is not a clan name.

The commonly accepted definition of an Aboriginal or Torres Strait Islander
is a person of Aboriginal or Torres Strait Islander descent who identifies as an
Aboriginal or Torres Strait Islander and is accepted as such by the community
in which he or she lives. The Torres Strait Islanders are the inhabitants of the
islands between Cape York and Papua New Guinea that were annexed to
Queensland in the 1870s. These peoples are mainly Melanesian, and their
culture and lifestyles are quite distinct from those of other Aboriginal people
(McRae, Nettheim and Beacroft, 1991: vii).

Aboriginal History: Phases in Government Public Policy

Most white Australians have grown up in ignorance of the Aboriginal
indigenous population of Australia. This ignorance is reflected in a lack of
understanding of Aboriginal traditions and culture, lack of knowledge of the
events and impact of colonial history and a scant appreciation of the situation
of Aboriginal people today (Freedman, 1989: 2). Foley describes a "conspiracy
of silence" which has allowed generations of Australians to grow up in ignorance
and to justify the oppression and exploitation of Aboriginal people (cited in
Roberts, 1978: ix).

In order to understand the position of Aboriginal people within Australian
society, it is necessary to provide an overview of the phases in official policies
which have occurred since colonization. The colonial background of Australia
has profoundly affected the dominant attitudes and assumptions of policy-

makers. Rowley (1970: 9) remarks:

> No adequate assessment of the Aboriginal predicament can be made
> so long as the historical dimension is lacking; it is the absence of
> information or background which has made it easy for intelligent
> persons in each successive generation to accept the stereotypes of an
> incompetent group.

Factors which contribute to the social, political and economic position of
Aboriginal people today have much of their basis in historical policies and
practices.   Although variations in policies and practices occurred across
Australia, there were many similarities and the actual outcomes for Aboriginal
people have been the same. These include a decline in population as a result
of disease and violence, the loss of control over the economic base, the loss of
control of sacred places and the destruction of native fauna and major
environmental changes resulting from the introduction of a materialistic culture
(Rowley, *op. cit.*: 355-366).   Middleton (1977: 81) comments that the
Aboriginal people were changed from "self-determining, semi-nomadic hunters
and gatherers into dependent, settled, unskilled labourers, held in subjection by
monopoly capitalism under conditions more often like slavery than wage
labour."

From the time of the arrival of Europeans in Australia, the continent has
been depicted as a new and pioneering frontier.   This myth has been
perpetuated through many avenues, including historical documentation, literature, the media and the development of a folk-lore surrounding white settlement
and rural development. The celebration of Australia's bi-centenary of European
settlement in 1988 reinforced such depictions and was considered offensive by
most Aboriginal people, many of whom protested against the celebrations.

Colonization

As Christie (1979: 118) suggests, the object of Great Britain in colonizing
Australia was far removed from the benefit and improvement of Aboriginal
people. Australia served initially as a dumping ground for convicts and then as
an outlet for Britain's teeming population and a source of raw materials for the
rapidly developing industries. Reynolds (1974: 308) refers to the overriding
commitment to material progress, which characterized Australia from the
earliest years of settlement.

Seizure of the Australian continent by the British occurred without any legal recognition of Aboriginal ownership of the land which was termed as waste and occupied. The prevailing international (European) law concerning the ownership of 'newly discovered' lands held that the inhabitants only had sovereignty over that land if, by their labour and practice of agriculture, they used it and changed it by constructing buildings and towns (Broome, 1982: 26). The policy in the early days of colonial rule has been described as a form of assimilation. Although not stated explicitly, it is clear that the colonizers hoped that Aboriginal people would be assimilated into the lower orders of the colonies and official statements of the time gave no recognition to the possibility that Aboriginal people might wish to retain their traditional way of life (Summers, 1975: 113). In 1835, Governor Gawler addressed Adelaide Aborigines as follows:

> Black men. We wish to make you happy but you cannot be happy unless you imitate white men. Build huts, wear clothes and be useful.... you cannot be happy unless you love God.... love white men.... learn to speak English.... (cited in Broome, *op. cit.*: 27).

Despite the 'civilize and Christianize' approach, the early days of colonization were characterized by genocidal practices. Aboriginal people were conquered by the force of arms, diseases to which they had no immunity, starvation and the destruction of their social systems (Grenfell Price, 1949). The desire to possess, to dominate and to colonize was at odds with humanitarianism (Broome, *op. cit.*: 27). The legal lie of *terra nullius*, a land belonging to no-one, opened the door to vigorous attempts at genocide (Secretariat of National Aboriginal and Islander Child Care, 1986).

Protection

It soon became evident that attempts to protect Aboriginal people from abuse and to 'civilize and Christianize' were unsuccessful, resulting in the development of protectionist policies throughout the continent. By 1911, all states, except Tasmania, had implemented special Aboriginal legislation with the emphasis on protection and restriction of Aboriginal people. The protection legislation of various states "expressed the determination to save, from abuse, people, who as experience was showing could not protect themselves. The Aborigine had proved himself a failure in society and could now be removed from it" (Rowley, *op. cit.*: 227). Significant in the development of protectionist policies was the belief that the indigenous population would die out in time.

With protection laws, Aboriginal people became a special category, subject to special laws for protection from the worst elements of contact with white society, while gradually 'civilizing' and converting to Christianity. Protectionist policies manifested themselves in various ways, but segregation and institutionalization of Aboriginal communities were in fact the main instruments (Gale, 1973: 58). Aboriginal people were segregated on reserves and missions without rights and responsibilities. Control was exercised over their employment, marriages, child care and education. Contact with whites only occurred for economic purposes and the land was left for total exploitation by whites (Evans, Saunders and Cronin, 1975: 121). Foley (1977: 5) describes the reserves and missions as "concentration camps."

In the short term, protection policies had advantages for whites in positions of power, as they removed Aboriginal people from the public gaze. The longterm consequences have been the production of a socially and economically deprived, depleted and dispossessed group, lacking in political power and basic human rights.

Assimilation

An assimilationist approach to Aboriginal policy emerged during the 1930s, but was not subject to any concerted implementation until after the Second World War. This policy was based on the notion of physical absorption for "half and lesser castes." The policy arose out of concern for some serious instances of brutality, murder and injustice directed at Aboriginal people, from support from missionaries and other groups for the idea of absorption for those who were "capable and suitable," pressure from humanitarian organizations and trade unionists and the effects of the 1930s depression on Aborigines. At the time this approach developed, the belief still existed that Aboriginal people were inferior. The Melbourne *Argus* newspaper in 1938 espoused the view that Aborigines were a "backward and lowly race" (Broome, *op. cit.*: 60). An obsession existed at the time with categorizing Aboriginal people according to their degree of Aboriginal heritage, reflecting the racist idea that the lighter the skin the more civilized and intelligent the person was supposed to be (*ibid..*: 61). Attempts were made to change Aboriginal people into Europeans with black skins (*ibid..*: 171).

Rowley (*op. cit.*: 331) comments that the objective of assimilation rocked no boats and was compatible with the continuance of the Aboriginal institutions which were being maintained by every government in Australia. The existing institutions changed from being instruments of protection to instruments of assimilation. Tatz (1972: 14) argues that for those on the receiving end, there was not much to choose between the paternalism of the assimilationist or segregationist variety. The policy has been described by Pittock (1977) as a subtle form of institutional racism.

The forced removal of children from their families was one of the most inhumane methods of implementing assimilationist policies, representing a significant means of exploitation by the State (Haukins, 1982: 1: 1: 2). Children were literally stolen from their parents and removed from their traditional way of life (Summers, 1975: 111). The separation of children from their families constituted a form of genocide, based on its philosophies of absorption, assimilation and ultimate disappearance of the Aborigine.

Self-management and Self-determination

The discrediting of assimilationist approaches eventually resulted in re-developed policies. A shift in policy emphasis emerged partly from a recognition that Aboriginal people might wish to retain their Aboriginal identity. Broome (*op. cit.*: 173) notes other influences, including increased non-British migration, the development of student movements, an alteration in church mission policies and media and United Nations attention on the plight of Aboriginal people. Attempts to move away from the assimilation approach have been translated into a number of policy phases, variously known as integration, self-management and self-determination.

A major policy breakthrough occurred following the election of the Federal Whitlam Labor Government in 1972. When the Labor Party came to power changes occurred on an unprecedented scale, with the new statements of policy acknowledging, for the first time, that white Australia owed a debt to those who had suffered conquest and occupation at its hands (Hamilton, 1974: 17). The initial optimism did not last, particularly as the new Government was forced to operate within existing structures and surrounded by the same paternalistic attitudes. A change to Conservative Government in 1975 reversed many of the policy initiatives.

A policy of self-determination, under a Federal Labor Government, remains in force at the present time. Despite this policy it is apparent that little has changed for Aboriginal people who are still subject to racism, discrimination, low socio-economic status and over-representation in welfare and legal systems. Some argue that assimilation policies remain pervasive. Undoubtedly, power remains in the hands of white society, with little or tokenistic Aboriginal participation in policy-making.

Other Significant Events

Citizenship Rights

In the 1940s, certificates granting Aboriginal people the rights of Australian citizenship were made available in most states to Aboriginal people who cared to apply for them and who met the conditions (Broome, *op. cit.*: 170). Certificates could be revoked, for example in Western Australia, for not altering to the habits of 'civilized' life. It appears that Aboriginal people largely rejected citizenship offers since it meant a severing of relationships with all Aboriginal people, other than immediate family (*ibid.*: 170-171). Reece (1979: 258) comments that even following the granting of citizenship, Aboriginal people were still subject to the discrimination of denial of voting rights and exclusion from the Census. The granting of citizenship can be viewed as a somewhat dubious honor, and certainly did not place Aboriginal people on an equal footing with white Australians.

The Referendum

Demands for Federal responsibility in Aboriginal affairs arose following pressure from Aboriginal organizations, the campaign for Gurindji land rights and the introduction of progressive legislation in South Australia. However, for this to occur it was necessary to remove the constitutional restrictions on Federal powers with respect to Aboriginal people. These constitutional provisions were:

> **Section 51**: The Parliament shall...have power to make laws for the peace, order and good government of the Commonwealth with respect to...the people of any race other than the Aboriginal race in any State for whom it is deemed necessary to make special laws.
> **Section 127**: In reckoning the numbers of the people of the Commonwealth or of a State or other part of the Commonwealth Aboriginal Natives shall not be counted.

It was considered that without the removal of the Constitutional impediment, the 'buck passing' between the Commonwealth (with money, but lacking authority) and the States (with powers, but insufficient funds) would continue. The referendum, held in 1967, received a record approval of 89.34 percent. This resulted in the removal altogether of Section 127, and Section 51 was changed by deletion of the words "other than the Aboriginal race in any State."

However, the Commonwealth decreed that as variations existed in the various states, administration had to occur on a regional or state basis. An office of Aboriginal Affairs was established within the Prime Minister's Department to co-ordinate and fund state initiatives.

Tatz (*op. cit.*: 101) is extremely critical of this decision as there were now seven major governmental administrative units for Aboriginal people as well as six basic sets of legislation and many variations on conditions for Aboriginal people. Criticism is still being levelled at the Commonwealth for its failure to exercise its constitutional authority, particularly in states where repression of Aboriginal people is in force. Crick (1981: 61) suggests that the referendum was an ambiguous event in the history of race relations in Australia as the Federal Government, despite its power, has been reluctant to use it.

Policy Influences

A number of factors have influenced the stages of policy development towards Aboriginal people by successive Australian governments. These include material gain for the British Empire and white settlers, the influence of race relations theories (particularly Darwinism) and the emergence of 'white nationalism.' More positive influences emerging since the Second World War have included increased contact with American blacks in Australia, the employ-ment of some Aboriginal people in the armed services, the struggle by Aboriginal pastoral workers for decent policies and conditions and an increase in contact with trade union organizations. During the post-war period there were significant changes in Australia's foreign policy and migration policy, combined with a growing sensitivity about a 'White Australia Policy' which restricted non-European migration to Australia. Such broad policy considera-tions had some impact on emergent policies towards Aboriginal people. Moreover, post-war employment opportunities resulted in the movement of many Aboriginal people to capital cities, bringing them closer to the center of political power and the media (Broome, *op. cit.*: 174).

It is evident that, despite changes of expressed policy, little has altered in practice and there have been minimal changes in the situation of Aboriginal people. Summers (*op. cit.*: 123) comments that the most striking feature of Aboriginal administration is the inability of governments to implement the policies they adopt. She suggests that they have been policies in name only and that the interests of the settlers and white landowners have taken precedence (*ibid.*). In respect to Victoria, Boas (1975: 11) argues that for more than a century after 1834, Aboriginal intervention remained a pragmatic process, guided by political pressures, first from England and then from pioneers, squatters and church missionaries. He comments that "Aboriginal intervention programs and legislation are seen primarily as the outcome of dilemmas usually created by a conflict of values" (*ibid.*). Tatz (1972) describes the introduction of new policies as innovations without change. Albrecht (1981) suggests that policies of self-determination and self-management have offered little because they are offered on white and not Aboriginal terms, and ignore Aboriginal structures and processes.

The National Aboriginal and Islander Health Organisation (undated: 2) comments that, despite stated policies of self-management and self-determination, policies of assimilation are continuing. The Organisation states:

> The difference from the pre-1972 policies being that they are now prepared to pay us to assimilate. We know however, that this policy will fail. If we do not succumb to bullets, poisoned flour and water holes, starvation, imprisonment and removal from our families, we will not succumb to Government money on Government terms (*ibid.*).

Tatz (1990) outlines twelve phases of attitude, belief or philosophy which have influenced government policy and administration related to Aboriginal people over the past 40 years. The phases are seen to be overlapping and somewhat inconsistent but they help to appreciate a give-and-take in policy development, noting that government policies are contingent on Aboriginal people's action and reaction, and vice versa. The stages range from the overtly racist phase to a land rights phase, through others which are characterized by Aboriginal militancy, victim blaming, focus on treaty making, recovery of rights and, lastly, community development. The inexorable move is to one where Aboriginal people are actors in determining their own future, not simply passive recipients of governments' largesse and sense of 'noblesse oblige.'

Comparisons with Other Indigenous Peoples

The history of treatment of Aboriginal people in Australia parallels that of other countries such as Canada. For example, the *Indian Act* in Canada (1876) provides a pretext for government interference in all aspects of the lives of indigenous people as well as a context for regulating all aspects of their lives and their communities. This Act is the instrument by which indigenous people were defined and re-defined over more than a century (for example, "Indian" is only a legal term, without reference to race or culture) and refers only to "registered Indians" who have a special legal status. The Canadian Federal Government only recognizes any legal obligation to registered Indians but nominally recognizes the ethnic group known as Metis (Frideres, 1988: 7). In this sense, more than 350,000 registered Indians are under the legislative and administrative (in)competence of the Federal Government and regulated by the *Indian Act*. In addition, there are in Canada more than 100,000 Metis, 35,000 Inuit (including Eskimo people living in the frozen north), and over 1,000,000 people of Indian ancestry. While treatment by successive Canadian governments toward Indian people have moved through similar phases as in Australia, the main differences between the two countries appear to lie in the existence of treaties which have established some bases for Indian claims regarding land rights. Treaty activity, while subject to differing expectations by government and Indian people, nonetheless proceeded, with the most significant date being October 7, 1763 when, by Royal Proclamation, the British Sovereign directed that all endeavors to clear the Indian title to land must be by Crown purchase (*ibid.*: 86). Claims processes, while slow and cumbersome, have remained a right rather than an exercise of charity. As well, recent government decisions in Canada, because of political uncertainties, constitutional crises and Indian self-determination (including armed struggle), are proceeding quickly to provide within five years a measure of self-government to Indian people within the Canadian Constitution and its institutions. While the nature of self-government has not yet been defined or well understood, it is clear that such government recognition of indigenous people's self-determination and their desire for self-government or community control in Canada is eons ahead of government thinking and action in Australia toward Aboriginal people, their rights and their aspirations.

The social and historical circumstances of Australian Aboriginal people are remarkably similar to those of the tribal people of India, the Indians of Mexico, Canada and American and the New Zealand Maoris (Lippman, 1970: 69). Lippman suggests:

> These and many other small groups of indigenes with a simple economy suffered physical defeat at the hands of more technologically advanced settlers. Their traditional lands, the basis of their economy, were purloined; they were pushed on to unproductive back country under a protectionist policy (which in practice was discriminatory) and were left, as a result, as a socially and economically depressed group, cut off from the rest of the community, suffering racial prejudice, still adhering in part to some aspects of traditional culture or to an adopted reservation culture, and with a strong sense of group identity (*ibid.*).

Aboriginal Disadvantage

The imposition of policies by a succession of white administrations on Aboriginal people continues to impact upon their condition to this day. Aboriginal people are on the lowest rung of the socio-economic ladder and morbidity and mortality rates far exceed those of the rest of the population. They are confronted with sub-standard housing, educational disadvantage and legal discrimination (Office of the Commissioner for Community Relations, 1981: 2). LaMacchia (1992: 14) notes that life amongst the chronically unemployed within Australia leads Aboriginal people to "wrong way dreaming — a merry-go-round of undereducation, unemployment, alienation and desperation."

According to the 1986 Australian Census of Population and Housing (in a country whose total population is approximately 17 million) the Aboriginal population is 227,645 (1.4 percent of the total population), of which 206,104 are Australian Aborigines and 21,541 Torres Strait Islanders (Australian Bureau of Statistics, 1987). By any standards, Aboriginal people are the most disadvantaged in Australia. Chief Justice Einfeld, for example, states that the living conditions of Aboriginal people in Australia may be likened to those in Soweto and Nazi concentration camps (Stevens, 1987: 4).

With regard to health status, Thomson (1991) states that Aborigines and Torres Strait Islanders comprise the least healthy identifiable sub-population in Australia and the standard of health Aboriginal people experience would not be tolerated if it existed in the population as a whole. While health problems vary across Aboriginal communities in the country and display some differences between those living in remote, less remote and urban areas, for almost all disease categories, rates are higher than for other Australians. For example, death rates are four times higher and life expectancy is up to 21 years less. While the causes of their lower health status are complex, social and economic inequality is considered to be central in importance.

The chronic use of alcohol and *kava* (a beverage made from the crushed root of the pepper plant), petrol sniffing and other substances such as analgesics and tobacco is considered to be high among Aborigines, although reliable statistics are hard to find since the extent to which Aboriginal people are identified in hospital and mortality statistics varies from state to state. Although Aboriginal people themselves have become more assertive in taking preventive and curative steps to deal with drug and alcohol abuse which have high social and physical costs, some studies have shown that there are structural issues of inequality and disadvantage which affect Aboriginal drinking patterns and the response of white people to this behavior. For example, Healy, Turpin and Hamilton (1985) examined Aboriginal drinking in a remote Queensland mining town, showing that restrictions to Aboriginal drinking were preserved through both overt and covert discrimination, making their drinking behavior more visible to the wider community and more accessible to police prosecution than that of any other ethnic group in town. They also showed that white people voiced unfounded fear about acts of violence and crime on the part of Aborigines, and the perpetuation of these attitudes and the resulting control measures taken against Aboriginal people themselves contributed to excessive consumption of alcohol by Aboriginal people.

The education system also disadvantages Aboriginal people to a marked degree. Cultural differences result in Aboriginal children approaching schooling and education with a radically different outlook to white children, with the dominant white culture causing severe disadvantage and feelings of low self-esteem (Australian Council of Churches, 1981: 43). As the Secretariat of National Aboriginal and Islander Child Care (*op. cit.*: 4) notes:

> The type of instruction to which the children are subject has been Eurocentric and assimilationist, contrived to fit them into the dominant Australian society. Aboriginal children have the lowest rate of participation in schooling. The acquisition of post-school qualifications is six times lower than for other Australians. Truancy is an extremely serious problem in Aboriginal communities.

In the child welfare field, statistics reveal a grim picture, including an over-representation of children as statutory clients of welfare departments, including as state wards and youth offenders. In New South Wales, as at November 1985, 12 percent of children in substitute care were Aboriginal although they comprised less than one percent of the total population of that State. In South Australia, as at August 1982, Aborigines represented 15-16 percent of all children under State care and control. In Western Australia, the Law Reform Commission noted that over 54 percent of children in foster care placements and over 58 percent of children in residential child care establishments were Aboriginal (1986: 235-236).

In the criminal justice area, Aboriginal people suffer constant indignities and disparate treatment. As a representative account of what happens to Aboriginal people in the justice system, a comprehensive study by Cunneen and Robb (1987) in New South Wales found that although Aboriginal people represented only 14 percent of the total population in the north-western area, they constituted 53.2 percent of all arrests and 52.7 percent of all court appearances during 1985-6. At the point of sentencing, Aborigines were almost four times as likely to receive a custodial sentence. The Law Reform Commission also suggested evidence of a link between a high rate of Aboriginal juveniles in corrective institutions and of Aborigines in prisons, and those persons having been placed in substitute care as children (*op. cit.*: 236). Evidence of this link has also been highlighted by Aboriginal organizations. The New South Wales Aboriginal Legal Service estimated that of the 525 Aborigines listed as being state wards in 1969, 50 percent had subsequently been in corrective institutions. An analysis of clients seeking assistance from the Victorian Aboriginal Legal Service for criminal charges, revealed that 90 percent of this group had been in fostering, institutional or adoptive placements (*ibid.*).

Prejudice, discrimination and violence toward Aboriginal people is an area of national disgrace. Irene Moss, Discrimination Commissioner of the Human Rights and Equal Opportunity Commission (1991), referring to the hearings of the National Inquiry into Racist Violence described below, states that:

> Of the many groups within our community affected by racist violence, Aboriginal people require special mention. The testimony and complaints that the inquiry received from Aborigines were quite disproportionate to their numbers in the Australian population. The evidence indicated that for Aborigines racist violence and harassment is a significant and almost universal experience in their daily lives. Aboriginal people who gave evidence to the inquiry made particularly serious charges about police.

Such evidence indicates that attempts by governments to improve the conditions of Aboriginal people have largely been a failure.

Recent Government Reports and Initiatives

Having noted the deplorable and inhumane conditions under which Aboriginal people live in Australia, and the daily experience of provocation, violence, prejudice and intimidation incurred by them, however, some initiatives have recently been taken by government, especially the Commonwealth government, to document and remedy injustices toward Aboriginal people. Whether these initiatives are sufficient in themselves or whether they contribute positively to some of the solutions required to affect Aboriginal disadvantage and bring some measure of equality is a topic of some controversy and debate in Australia.

Racist Violence

The Human Rights and Equal Opportunity Commission in 1990 conducted an Inquiry into Racist Violence in Australia. Although the inquiry was precipitated by a number of racist incidents involving both migrants and Aboriginal people and by the perception of a growing racism and prejudice in the country, the Report highlights despicable treatment of Aboriginal people, including violence against Aboriginal organizations and individuals, in public places such as parks, streets and hotels, affecting their cultural and social life. In fact, many of the complaints (in 25 of 50 written submissions to the Inquiry relating to Aboriginal and Islander people of racist violence) had to do with police behavior, mainly unprovoked and irrational, toward Aboriginal people. Some of the most common complaints made against police were the threat of force used in making arrests, the use or threat of physical violence while in custody (particularly with juveniles), and the provocative and disrespectful language and rough handling of Aboriginal women by male police officers (National Inquiry into Racist Violence, 1991: 80).

Deaths in Custody

In 1991 the Royal Commission into Aboriginal Deaths in Custody reported its investigation into the deaths of 99 Aboriginals and Torres Strait Islanders in police custody or in youth detention institutions between 1980-87. The Commission's investigation, while not finding any deaths to be the result of deliberate unlawful violence or brutality by police or prison officers, stands as an indictment of the legal and corrective services system as it is applied to the most disadvantaged group in Australian society. In a classic understatement, the Commonwealth Minister for Aboriginal Affairs as he released the Report noted that the Commissioners found that "there has been a failure to live up to the standard of care owed to those in custody.... It has found many system defects and many individual failures to exercise proper care and basic human compassion." The Commonwealth, State and Territory governments have responded to the Report with a lack of alacrity, while a *Deaths in Custody Newsletter* (Searcy, January 1993) reports that nationally there have been at least 28 Aboriginal deaths in custody since 1988. One of the most infamous among the facilities is the Townsville Correctional Center, of which the Human Rights Commissioner claims that the conditions in the gaol are in breach of Australia's human rights obligations (*ibid.*).

In drawing up a profile of those who died, it was found that:

...of the ninety-nine, eighty-three were unemployed at the date of last detention; they were undeducated — at least in the European sense — or under-educated — only two had completed secondary level; forty-three of them experienced childhood separation from their natural families through intervention by the State authorities, missions or other institutions; forty-three had been charged with an offence at or before aged fifteen and seventy-four had been taken into last custody directly for reasons related to alcohol and it can safely be said that overwhelmingly in the remaining cases the reason for last custody was directly alcohol related.... generally speaking the standard of health of the ninety-nine varied from poor to very bad (the average age of those who died from natural causes was a little over thirty years); their economic position was disastrous and their social position at the margin of society; they misused alcohol to a grave extent.... (Royal Commission into Aboriginal Deaths in Custody, 1991: 5-6).

It was found that as well as having early contact with the criminal justice system, Aboriginal people had repeated contact with it. The Commission established that Aboriginal people in custody do not die at a greater rate than non-Aboriginal people in custody. However, what is markedly different is the rate at which Aboriginal people come into custody compared with the rate of the general community. The degree of over-representation in police custody, as measured by the Commission's study of police cell custody in August, 1988, is twenty-nine times (*ibid.*: 6).

In the Report, reference is made to the importance of history, arguing that from that history many things flow which are of central importance to the issue of Aboriginal over-representation in custody. These include the systematic disempowerment of Aboriginal people, dependence on government and decisions made about them and imposed on them (*ibid.*: 8). History also had an impact on non-Aboriginal people, as every turn in the policy of government and the practice of the non-Aboriginal community was postulated on the inferiority of the Aboriginal people (*ibid.*: 9).

The Commissioner also highlighted the fact that relations between white and black people have been marked by inequality and control. The relationship was at its worst between Aboriginal people and the police forces of the dominant society (*ibid.*: 10).

The 339 recommendations from the Report are wide-ranging, including the elimination of abuse of Aboriginal persons by police officers, the introduction of measures to reduce the rate at which Aboriginal juveniles are involved in the welfare and criminal justice systems, the abolition of the offense of public drunkenness, the introduction of legislation to enforce the principle that imprisonment should be utilized only as a sanction of last resort and placement of Aboriginal prisoners in an institution as close as possible to his or her family (*ibid.*: 31-108).

Early response to the Royal Commission's Report by governments has been positive, with a recognition that over-representation and the unacceptable number of deaths in custody are due to the disadvantaged and unequal position of Aboriginal people in the society — socially, economically and culturally — and an acceptance of most of its recommendations.

> We therefore fully accept the Royal Commission's insistence that the problem be tackled simultaneously at two levels — an improvement to the position of Aboriginal and Torres Strait Islander people in relation to the criminal justice system, and a strengthened commitment to correct the fundamental factors which bring Aboriginal and Torres Strait Islander people into contact with that system (Overview, 1992: 1).

New Forms of Governance

The Commonwealth Government has, in the past decade, also undertaken a number of structural initiatives to encourage and support Aboriginal participation in assuming more responsibility and self-governance in relation to land and other areas.

The major structural change was to establish the Aboriginal and Torres Strait Islander Commission (ATSIC). Until March 1990 indigenous affairs was mainly administered through the Commonwealth Department of Aboriginal Affairs. In November 1989 legislation was passed which brought about the amalgamation of this Department and the Aboriginal Development Commission into the new Commission. ATSIC represents an attempt to give Aborigines and Torres Strait Islanders the power to make decisions about programs that affect them. Sixty Regional Councils have been established throughout Australia, comprising 790 Regional Councillors. All are Aboriginal or Torres Strait Islanders.

Another recent initiative was to establish an Aboriginal Reconciliation Council which emanates from a commitment by the Commonwealth Government to a process of reconciliation between Aboriginal and Torres Strait Islander people and the wider community. The purpose of the reconciliation process is the transformation of Aboriginal and non-Aboriginal community relations in Australia. The process of reconciliation was endorsed by the Royal Commission into Aboriginal Deaths in Custody as a means of avoiding community division, discord and injustice to Aboriginal people. The final report of the Commission notes that Australia has to recognise that "there was a large reservoir of distrust, enmity and anger among Aboriginal people and a lack of understanding among non-Aboriginal people" (Forbes, 1993: 17). The reconciliation process has three main components:

(1)   An ongoing public awareness initiative to educate non-Aboriginal people about Aboriginal history, culture, dispossession, continuing disadvantage and the need to address that disadvantage.

(2)   The fostering of a national commitment at all levels to co-operate in addressing the land, housing, health, law and justice, education, employment, infrastructure and economic development needs of Aborigines and Torres Strait Islanders.

(3)   Extensive community consultations with Aborigines and Torres Strait Islanders and non-Aborigines about whether reconciliation would be advanced by a formal document or instrument of reconciliation, or the nature of that instrument and the process for concluding it.

The process is to be guided and advanced over a ten-year period by a Council for Aboriginal Reconciliation composed of a maximum of 25 members, with approximately half of them indigenous people. The legislation to establish the Council places considerable emphasis on action at the community level.

Aboriginal Views and Thrusts

One myth, rapidly being reduced to the rubbish bin, is that Aboriginal people have been and still are passive and dependent 'no-hopers.' Speaking of moves by Aborigines to take more direct control over their own affairs, Bird and O'Malley (1987: 48) state:

> To achieve actual decolonialisation Kooris require land rights, a secure economic base, and control of the bureaucracy which 'serves' them. The push to make this a reality is coming on a number of fronts as a result of Koori initiatives in international forums, in Australian courts of law, and through direct and indirect political action. In all

these arenas Kooris are eschewing welfarist solutions and are increasingly framing their discourse in terms of their continuing 'sovereignty.' By constantly articulating their needs at the structural level, Kooris are in the vanguard of attempts in Australia to achieve a meaningful measure of social justice.

Community Control

The concept of community control is central to Aboriginal demands. A statement by the World Council of Churches captures the spirit of the Aboriginal quest for autonomy and self-determination:

> What we have heard is that Aboriginal communities want to recover their human dignity and respect so that they can break their chains of dependence, alienation from their culture and recover their history and dignity. They want their self-development through social justice, self-reliance and economic well-being and not through permanent dependency as the price for survival, and alienation from their people and history as a price for decent housing, proper medical facilities and white Australian educational models. People want to be free to be human with the freedom to say 'no' and the dignity to make their own mistakes (Australian Council of Churches, 1981: 11-12).

The Victorian Aboriginal Task Force on Land Rights and Compensation (undated) states:

> Aborigines have been forced to live in their own country under policies of restriction, assimilation and slavery. There has been no benefit for them there. True self-determination allows Aborigines to determine their own destiny — a right of all people.

Alongside this key Aboriginal demand is the quest for adequate resource provision, cultural recognition and consultation which are seen as essential corollaries to community control.

It is evident that directions towards community control are unable to proceed satisfactorily without adequate provision of resources. Inadequate funding frequently results in accusations by Aboriginal organizations that governments 'set up programs to fail.' Fear of a white backlash, the influence of vested interest groups, including the mining lobby, and the emergence of 'new right' philosophies in Australian society impact on government resistance.

Chisholm (1985: 113) argues that reforms offered to Aboriginal people offer a "precarious kind of self-determination, one that depends on continued acquiescence of government authorities." Broome (*op. cit.*: 199) comments that Aboriginal organizations in their quest for funding are often at the mercy of insensitive and distant government bureaucrats.

Aboriginal people have strongly resisted attempts to be absorbed and assimilated into mainstream Australian society and have expressed the wish to maintain their own culture, identity and traditions in all aspects of their lives. Questioned is the ability of white Australians to understand the nature of the Aboriginal way of life, and the belief exists that maintenance of culture can only occur through Aboriginal-controlled mechanisms. As noted in the Report of the Royal Commission into Aboriginal Deaths in Custody (1991), Aboriginal people:

> never voluntarily surrendered their culture and, indeed, fought tooth and nail to preserve it, throughout dispossession, protection, assimilation, integration. In their own words, they survived and their culture survived; in different forms and to different degrees in different parts of the country as a result of different experiences. They have the right to retain that culture, and that identity. Self-determination is both the expression and the guarantee of that right (19-20).

Another common criticism of government by Aboriginal people relates to their failure to consult with Aboriginal communities and their failure to ensure Aboriginal participation in the decision-making process. Australia has a long history of imposition of policies and programs upon Aboriginal people with inadequate or tokenistic consultation. Aboriginal organizations and communities operate on a consensus and participatory approach to decision-making which governments frequently fail to recognize. Aboriginal people are critical that decisions are frequently left to educated whites who have no knowledge of Aboriginal communities and their people, and little consideration is given to the expertise of Aborigines. The Royal Commission into Aboriginal Deaths in Custody has noted the need for Aboriginal participation in decisions affecting their lives (Muirhead, 1988: 65).

The major mechanism employed by Aboriginal people to wrest control from governments is the formation of Aboriginal organizations. As noted by Bodley (1986: 158), "indigenous people are designing power structures that permit the consolidation of a power base to successfully confront states without sacrificing their egalitarian and communal characteristics." He notes that the problems confronting indigenous peoples are political power problems, with indigenous peoples being destroyed because they lack the political power to adequately defend themselves against dominant societies and to press for their demands (*ibid.*). From early this century, Aboriginal people have endeavored, through the establishment of such organizations, to have their plight recognised and redressed. A large number of specialist organizations exist throughout Australia including in the fields of health, education, child welfare, legal services and land rights.

Aboriginal thrusts represent a movement away from welfare responses. Bird and O'Malley argue that the "welfarist" model of social justice is based on individual responses to disadvantage, with policies and programs designed to overcome the effects on an individual of bad luck or bad management (1987: 45). Capitalist political-legal systems are based on the individual, whereas Aboriginal society, being kin-based, requires community control of resources and organizations (*ibid.*).

Land Rights

Perhaps the biggest clash between white and black people throughout Australia's post-colonial history has related to the land. Aboriginal claims for land rights have questioned the legal assumptions on which two centuries of land ownership and use have been based (Rowley, 1986: 68). They also represent a major expression by Aboriginal people to gain control over their lives and their heritage.

More fundamentally, the quest for land rights demonstrates the difference between 'white' and 'black' values in relation to land. The Australian Council of Churches has described this difference as one example of the "error of materialism" whereby directly or indirectly, the spiritual and personal is subordinate to material reality. The cultural meaning of land is limited to its physical or material usefulness (1987: 5).

The Chairperson of the Northern Land Council, Galarrwuy Yunupingu states:

> For Aboriginal people, there is literally no life without the land. The land is where our ancestors came from in the Dreamtime, and it is where we shall return. The land binds our fathers, ourselves and our children together. If we lose our land, we have literally lost our lives and spirits and no amount of social welfare and compensation can ever make it up to us (cited in Department of Aboriginal Affairs, 1984(ii): 2).

The struggle by Aboriginal people for land rights in Australia has had a long and shaky history. As recently as 1992, Yunupingu asserted that:

> They don't poison our water holes any more in the Northern Territory, but the Territory Government here — the Country Liberal Party mob — break their agreements to grant any Aboriginal people any excisions or living areas, and they always take us to court to hold up our legitimate land claims (cited in *The Age*, 2 April 1992).

A breakthrough occurred in an historic judgement, known as the Mabo case, which "swept the concept of terra nullius into the trash can of history, where it belongs, and has recognised native title..." (Forbes, 1993: 17).

Eddie Mabo was a member of the Meriam people, the traditional owners of the Murray Island and surrounding islands and reefs in the Torres Strait. The islands in the strait were annexed as part of the colony of Queensland in 1879. In 1982, Mabo and four other Islanders commenced action in the High Court of Australia seeking a declaration of their traditional land rights, claiming that the islands had been continuously inhabited and exclusively possessed by their people (Brennan, 1992: 1). The State of Queensland attempted to defeat the claim by introducing the *Queensland Coast Islands Declaratory Act 1985*, which declared that, upon the islands being annexed, they were vested in the Crown in the right of Queensland. In 1988, the High Court ruled this Act was contrary to the Commonwealth's *Racial Discrimination Act* and in 1992 upheld the Islanders' claim, ruling by six to one that the lands of the continent were not *terra nullius* in 1788 (*ibid.*).

Three judges involved in the Mabo case stated that it was imperative in today's world "that the common law should not be seen to be frozen in an age of racial discrimination" (*ibid.*). The nervousness of governments about the implications of the ruling for other land claims is fast becoming apparent.

Focus on Aboriginal People in the Justice System

As noted briefly above, Aboriginal people are not adequately served by the justice system in Australia. Control over them is enforced by the workings of the state apparatus, including governments, the police and the courts as state agents. Although subject to the same laws and having the same rights and obligations as other Australian citizens, Aboriginal people are arrested, charged, convicted and imprisoned at rates many times in excess of the rest of the Australian community (Aboriginal Australia: 1).

In terms of the national situation, research by the Australian Institute of Criminology has found that the imprisonment rate of Aboriginals is 23 times higher than for white Australians (Burchell, 1987: 17). Figures released by the Institute in 1988 showed an increase in the proportion of Aborigines in prison from 13.4 in 1985 to 14.5 in 1986 (Metherell, 1988: 5). Aboriginals are imprisoned largely as a result of offenses against the person, theft and motor vehicle offenses, with the inability to pay fines being a major contributor to high imprisonment rates. Drunkenness and alcohol-related offenses are also seen to have played a major role in the imprisonment of Aboriginal people (Graham, 1989; Langton, 1988).

Statistics on Aboriginal over-representation in state correctional systems tell the same story. For example, an examination of statistics from Victoria (Victorian Office of Corrections, 1990: 1-4) reveals that:

(1)   On a per capita basis, Aborigines are imprisoned almost 13 times as frequently as all Victorians.

(2)   The Aboriginal imprisonment rate increased by around 95 percent between 1986 and 1989.

(3)   Aboriginal prisoners are predominantly male, young, single, poorly educated and unemployed, to an extent more accentuated than in the general prison population.

(4)   Aboriginal prisoners tend to be serving shorter sentences than other prisoners.

(5)   Forty-three percent are serving aggregate sentences of less than two years.

(6)   On a per capita basis, Aboriginal people are eight times more likely to be serving a community-based disposition than non-Aboriginal Victorians.

Initial Contacts With the System

Aboriginal people, in their first contacts with the justice system, are treated differently, for whatever reasons (which may be racial), than other Australians. This imbalance is initially reflected in the high proportion of Aboriginal people who are directed to the courts rather than being given warnings at the time of apprehension (Gale and Wundersitz, 1986). For example, in a case study in South Australia, it was seen that there are three crucial points within the juvenile justice system which involve the exercise of individual discretion: police decisions whether to initiate criminal proceedings and whether to apprehend by way of an arrest or report; the referral decisions of Screening Panels; and Children's Court sentences. Aboriginal youth are subject to harsher treatment options at all stages of the criminal process than non-Aboriginal youth. They are substantially more likely to be apprehended and brought into the system than are members of other ethnic groups studied; once the decision to initiate criminal proceedings has been taken, Aborigines are more likely to be arrested than reported by police; to be referred to the Children's Court rather than diverted to Children's Aid Panels; and once before the Children's Court to be sentenced to detention. According to the 1986 census, while Aborigines represented only 1.7 percent of the total youth population of that state, for the year ending 30 June 1986 they accounted for 9.3 percent of all appearances before Children's Aid Panels (non-judicial and informally structured panel) and the Children's Court. Screening panels are far more likely to send Aboriginal youth to Court rather than divert them to Aid Panels which are less stigmatizing and do not impose conventional penalties. (Gale, Bailey-Harris and Wundersitz: 41)

Disparities in Sentencing

As noted above, it is estimated that Aborigines are 23 times more likely to be gaoled than non-aboriginal people. McDonald (1982) observes that Aborigines are less likely to receive bail or to be represented by a lawyer, despite the existence of the Aboriginal Legal Service. However, even when bail or probation is received, magistrates often use methods of retribution such as banishment, which is unrealistic in a social and cultural sense for Aboriginal people and makes further criminalization more likely (National Inquiry Into Racist Violence: 1991). The disadvantage of Aboriginal juveniles is compounded as they move through the systems as their chances of being imprisoned are twenty-three times greater than their population size would suggest. It was also found that while Aborigines are generally more likely to be denied the opportunity of diversion to an Aid Panel, "it would appear that only the operation to some degree of racial bias could account for these inconsistent applications" (Gale and Wundersitz, *op. cit.*: 91). Meanwhile, McRae, Nettheim and Beacroft (1991) note that Aborigines are also more likely to plead guilty, and more likely to receive a prison sentence than a non-custodial sentence. On the other hand, sentences for Aborigines tend to be shorter, although this may reflect the nature of the offenses for which they were incarcerated, as they tend to be of a less serious nature as we have seen above.

The Custodial System

'Prison' seems like such an innocuous word until the inhumanity of prisons is uncovered. For example, an Amnesty International report has described the conditions of some of the detention facilities in which Aboriginal people are held as being cruel, inhuman and degrading. It emphasizes that the criminal justice system makes Aborigines more likely than other Australians to suffer imprisonment and to have their rights violated (*The Age*, 11 February 1993). The indignities suffered by Aboriginal people in prisons are often preceded by their disdainful treatment during the interrogation process where they are victims of a lack of knowledge of their rights, intimidation by police and violent beatings (Graham, 1989; Kearins, 1991). Complaints about treatment within custody, as documented by the Royal Commission on Deaths in Custody, range from racist abuse to denial of proper medical treatment, hosing down, harassment and serious overcrowding in cells. The causes of deaths in custody are debated by Aboriginal activists and the police (Wooten, 1991), and whether they are primarily suicide or murder (Graham, 1989; Wooten, 1991; Biles, 1991).

A Whitefellas Justice System

The nature of police-Aboriginal relations must, however, be noted as a salient factor to examine in explaining the over-representation of Aboriginal people in the justice system of Australia. LaMacchia (*op. cit..*) describes this relationship as one of "mutual hostility," in which police are seen by Aboriginal people as "armies of occupation," agents of a "whitefellas system" which has no legitimacy and is seen as alien, imposed and essentially irrelevant to Aboriginal people. The perception is one where Aboriginal people are persecuted by a foreign white culture which persecutes them for behavior which is either acceptable in their own culture or is a response to the influence of the white culture (McCorquodale, 1987).

Aboriginal people view this system which controls them to be simply a 'system of injustice' imposed on them by the occupiers of their traditional lands, by people who do not understand and who do not want to understand them and their culture. Aboriginal people seldom have any input into correctional services which are meant to serve them (Davidson, 1988). Police attitudes and practices appear to be a product of the society which supports existing inequality and discrimination (Law Reform Commission, 1986). These attitudes may be more easily identified in a police force whose officers act as 'front-line workers' representing a white society from which the officers are derived. However, we note that a number of community forums have raised demands by white community groups for stricter police presence in Aboriginal areas in order to protect "white" citizens' homes and businesses.

Discriminatory police practices and deficiencies in police knowledge and understanding of Aboriginal people, their culture, values and their lifestyles seem to be a salient factor in the maintenance of a mutual hostility. For example, in the Redfern area of Sydney in the late 1960s and early 1970s, there were complaints from Aboriginal people that they were being arrested without cause and were the subject of a police-imposed curfew (Lyons: 138), while complaints concerning police activities have continued into the nineties (Cunneen, 1990: 8). One of the most blatant abuses of police power was exhibited in the infamous policing operation referred to as the "Redfern Raid of 8 February 1990." The operation, code-named "Operation Sue," included 135 police from a Tactical Response Group, the Anti-Theft Squad, the Rescue Squad and other officers. Some 70 police raided eight to ten houses, with the remainder of the police providing back-up support and sealing off the area. Armed with at least eight search warrants, and using iron bars and sledgehammers to gain entry (with some police carrying firearms and batons), only eight arrests were made, one of those a juvenile. With such a massive use of force by police and with the full weight of the justice system behind them, three persons were charged with goods in custody, one person, a juvenile, was charged with possession of an implement for the use of drugs, two persons were detained for breach of bail and failure to appear (and another warrant for being under the influence of alcohol on the railways). Despite the fact that excessive force had been used to gain entry and after entering the houses, and that the houses were left in poor condition after the raid, the Chief Inspector and the Minister of Correctional Services stated that the operation '"was mounted to protect the law-abiding citizens of the area who are, of course, the vast majority," with the implication that those who criticized the raid were somehow not part of the law-abiding majority in the Aboriginal community. (*ibid.*: 23). A Report commissioned by the Human Rights and Equal Opportunity Commission concluded that on the evidence available, overpolicing, including the excessive use of force, occurs in Redfern and that it is open to find that the "Redfern Raid" constituted an act of racist violence, within the terms of reference of the National Inquiry into Racist Violence (*ibid.*: 34-35). It is also found that avenues for complaints against police conduct are threatening for Aboriginal people; they express concerns about abuse and harassment if they lodge such complaints (Sculthorpe, 1990).

Some Examples of New Initiatives Required in the Justice System

The justice system in Australia, heavily weighted to disadvantage Aboriginal people, needs to undergo radical and comprehensive changes. Some of the changes are attitudinal, others are structural and institutional. The system has to be resposive to Aboriginal people, their culture, their needs and aspirations, and their communities. It has to be cognizant of the need for greater community control by Aboriginal people, as well as the necessity to divert Aboriginal people from the justice system itself. Within the justice system, new initiatives such as the use of community panels and customary law are a requisite for achieving justice. We will outline a few of these initiatives, recognizing that they are but a few pieces to the puzzle.

Community Panels

One of the outcomes of the Royal Commission's inquiry into Aboriginal deaths in custody has been the establishment of Community Justice Panels (CJPs) in Aboriginal communities throughout Victoria. The Panels comprise a group of representatives chosen by their communities to liaise with the local police, the courts and the jails, operating like Elders' Councils (Victorian Aboriginal Legal Service, 1989: 29). The program is designed "to prevent the killing of the Koori by the Gubba criminal justice system" (*ibid.*). *Gubba* is a term frequently used by Aborigines to refer to white people.

The program provides an example of a community-controlled initiative which operates through existing structures. Located within the Victorian Police Force, Aboriginal people, through meetings of the regional panel chairpersons, dictate the terms of the program to the police.

The Panels aim to limit the time Aboriginal people spend in formal contact with the judicial system and in police custody. Volunteer panel members, who are on call 24 hours a day, are contacted by the police when an Aboriginal person is apprehended. Preferred methods of diversion include admitting the person to a "sobering up center," sitting with them to assist in "cooling off" or taking them home.

Additional benefits arising from the scheme have included better relations between police and Aboriginal communities, the halving of police/Aboriginal contact in some locations and the establishment of additional programs by some CJPs to respond to their community needs. Examples of these programs are crime prevention schemes, cultural camps for young people and a prison visiting service.

The program is consistent with the Royal Commission recommendation that the police services should adopt and apply the principle of arrest being the sanction of last resort (Royal Commission into Aboriginal Deaths in Custody, *op. cit.*: 50).

Customary Laws

In 1986 a report into an inquiry into the recognition of Aboriginal customary laws by the Australian Law Reform Commission was tabled in the Federal Parliament. The report contained extensive discussion of the relationship between Aborigines and the criminal justice system. One of the issues the Commission was asked to consider was "whether and in what manner existing courts dealing with criminal charges against Aborigines should be empowered to apply Aboriginal customary law and practices in the trial and punishment of Aborigines" (Hennessy, 1988: 13).

The Commission's Report sets out the basic principle that the Australian criminal law should apply to all persons, Aboriginal and non-Aboriginal. However, in its application to Aboriginal people special consideration needs to be given to the continued operation of Aboriginal customary laws. The Commission specified a number of sentencing principles for taking account of Aboriginal customary laws (*ibid.*). The Report was directed at the recognition of Aboriginal customary laws and not the more general issue of Aboriginal involvement with the Australian justice system. It recommended that customary law should be recognised by the criminal justice system. The most appropriate way for this to be done was at the sentencing stage of the process, rather than by creating new offenses or defenses (*ibid.*: 14).

Three years after the tabling of the Report, the then Federal Minister for Aboriginal Affairs reported that there had been gradual but steady progress towards implementing recommendations, whilst noting implementation difficulties. One issue raised was that most recommendations would be administered by state and territory governments and the legislative response to recommendations could vary in each jurisdiction (Aboriginal Law Bulletin, 1989: 15).

In summary, there are a variety of ways to make changes to provide more justice for Aboriginal people in the justice system. Some of these avenues include efforts to:

(1)   'indigenize' the justice system by introducing more customary law, by hiring more Aboriginal police, by using Aboriginal

Justice Panels, Aboriginal court workers and other means;

(2)   by changes to the justice system, which discriminates against Aboriginal people, by training judges, magistrates, lawyers and police in Aboriginal history and customs, by improving police-Aboriginal community relations by training police at all levels in cross-cultural relations and access and equity, by introducing a range of dispute resolution mechanisms, by use of mechanisms   and programs which divert Aboriginal people from the prevailing justice system, and by other means such as attention to bail, detention, and other penalties such as community service orders; and

(3)   by granting Aboriginal people more control over their own governance, so that it is 'their' systems which prevail and not those of 'whitefellas.'

Ultimately, the best provisions will be those which reduce the contact between Aboriginal people and the processes of law, reducing the stigmatization, brutality, inhumanities and injustice which they suffer and endure. This means attention to improving the conditions in which Aboriginal people live and resolving the economic and social problems they face in this country. In this way, one will not have to ask the question whether Aboriginal people are receiving justice or differential treatment.

Conclusion

Aboriginal people in 1993 remain the most disadvantaged people in Australia, with many living a kind of 'fourth-world' existence characterized by poverty, ill-health, unemployment, bad housing and discrimination. Despite recent government (particularly Commonwealth) actions to remedy Aboriginal disadvantage and to seek reconciliation of the dominant society with Aborigines, the remedies taken or planned do not appear to be sufficient to allow indigenous people in Australia to be full citizens of, and achieve real equality within Australia. Good intentions do not equate with justice and equality, and government initiatives are seen by many Aboriginal people as 'too little, too late.' The much vaunted Aboriginal Reconciliation Council is not yet tested as an effective instrument by which the dominant society and the Aboriginal community can become reconciled within this vast continent. Aboriginal people need to have their land rights respected and to have more control of their economic base and perhaps their own governance. There is as yet only a 'hollow hope' that Aboriginal people will significantly move toward full equality within Australia in 1993.

References

Aboriginal Australia. 1989. *Current Issues: Law and Justice*. Canberra: Department of Aboriginal Affairs.

Aboriginal Law Bulletin. October 1989. Number 40.

Albrecht, Paul G.E. December 1981. "Talking on Whose Terms," *The Bulletin*.

Australian Bureau of Statistics. 1987. *Aborigines and Torres Strait Islanders, Australia, states and territories*. Catalogue no. 2499.0, ABS. Canberra.

Australian Council of Churches. 1981. *Justice for Aboriginal Australians: Report of the World Council of Churches team visit to the Aborigines*. Sydney.

Australian Council of Churches. 1987. *A Just and Proper Settlement*. Sydney.

Australian Law Reform Commission. 1986. *Recognition of Aboriginal Customary Laws*. Sydney: ALRC.

Barnard, J. 1975. "Aborigines of Tasmania," cited in Evans, R., K. Saunders, and K. Cronin, *Exclusion, Exploitation and Determination*. Sydney: Australia and New Zealand Book Co.

Basten, John, Mark Richardson, Chris Ronalds, and George Zdenkowski (eds.). 1982. *The Criminal Injustice System*. Sydney: ALWG (NSW) and Legal Service Bulletin.

Biles, D. 1991. "Deaths in Custody: The Nature and the Scope of the Problem," *Criminology Australia*. May.

Bird, Greta and Pat O'Malley. 1987. "Kooris, Internal Colonialism and Social Justice," *Social Justice*. Special Edition on The Politics of Empowerment in Australia, Volume 16, number 3, pp. 35-50.

Boas, Phillip J. 1975. "A Case Study in the Practice of Social Intervention in Aboriginal Affairs (Victoria) 1834-1972." Unpublished thesis. University of Melbourne.

Bodley, John H. 1986. *Victims of Progress*. Sydney: ALRC.

Brennan, Frank. 1992. *Mabo and its Implications*. Canberra: Uniya.

Broome, Richard. 1982. *Aboriginal Australians*. Sydney: George Allen and Unwin.

Burchell, David. December 1987. "The Bicentennial Dilemma," *Australian Society*.

Chisholm, Richard 1985. *Black Children: White Welfare?* Sydney: Social Welfare Research Center.

Choo, Christine. 1990. *Aboriginal Child Poverty*. Melbourne: Brotherhood of St. Laurence.

Christie, M.F. 1979. *Aborigines in Colonial Victoria 1835-86*. Sydney: University of Sydney Press.

Crick, Malcolm. 1981. "Aboriginal Self-management Organizations: Cultural Identity and the Modification of Exchange," *Canberra Anthropology*. Volume 4.

Cunneen, Chris. November 1991. "Aboriginal Juveniles in Custody," *Current Issues in Criminal Justice*. Volume 2, number 2, pp.204-218.

Cunneen, Chris and Robb, T. 1987. *Criminal Justice in North-West N.S.W.* Sydney: N.S.W. Bureau of Crime Statistics and Research, Attorney General's Department.

Cunneen, Chris. May 1990. *Aboriginal-Police Relations in Redfern: With Special Reference to The 'Police Raid' of 8 February 1990*. Report Commissioned by the National Inquiry into Racist Violence. Sydney: HREOC.

D'Souza, Nigel. June 1990. "Aboriginal children and the juvenile justice system," *Aboriginal Law Bulletin*. Volume 2, number 44, pp. 4-5.

Davidson, Don. 1988. "The Aboriginal and Torres Strait Islander viewpoint into corrective services in Queensland," *Justice, Law and Prisons Seminar*. Brisbane: Q.E.A. Legal Service.

Department of Aboriginal Affairs. 1984. *Current Issues in Aboriginal Affairs*. Canberra: Australian Government Publishing Service.

Department of Aboriginal Affairs. April 1984 ii. *The History of Aboriginal Land Rights in Australia*. Canberra: DAA.

Department of Aboriginal Affairs. April 1989. *Aboriginal Australia: Current Issues, Law and Justice*. Canberra: DAA.

Dodson, Mick. October 1987. "Aborigines and the Criminal Justice System," *Aboriginal Law Bulletin*. Number 28.

Evans, R., K. Saunders and K. Cronin. 1975. *Exclusion, Exploitation and Extermination*. Sydney: Australia and New Zealand Book Company.

Fesl, Eve Mungwa D. Fall 1987. "White Manoeuvers and Koori Oppression," *Social Justice*. Special Edition on The Politics of Empowerment in Australia. Volume 16, number 3, pp. 30-34.

Foley, Gary. 1977. "Blacks for Australian Independence," *Aboriginal and Islander Identity*. Volume 3, number 3.

Forbes, Cameron. 1993. "Black Horizon," *The Weekend Australian*, February 20-21.

Freedman, Linda. 1989. "The Pursuit of Aboriginal Control of Child Welfare." Unpublished M.S.W. Thesis. The University of Melbourne.

Frideres, James S. 1988. *Native Peoples in Canada: Contemporary Conflicts*. Scarborough: Prentice-Hall Canada.

Gale, F. 1973. *Urban Aborigines*. Canberra: A.N.U. Press.

Gale, Fay and Joy Wundersitz. March 1986. "Aboriginal Visibility in the System," *Australian Social Work*. Volume 39, number 1.

Gale, Fay, Rebecca Bailey-Harris and Joy Wundersitz. 1990. *Aboriginal youth and the criminal justice system: the injustice of justice?* Cambridge: Cambridge University Press.

Graham, D. 1989. *Dying Inside*. Melbourne: George Allen and Unwin.

Grenfell Price, A. 1949. *White Settlers and Native Peoples*. Melbourne: Georgian House.

Hamilton, Annette. 1974. "Aboriginal Cultures. Management or Autonomy," *Arena*. Number 34.

Hanks, P. and B. Keon-Cohen, B. (eds.). 1984. *Aborigines and the Law*. Sydney: Allen and Unwin.

Haupt, Robert. November 1987. "The Aboriginal Condition: The Brute Facts," *The Age*.

Hazelhurst, Kayleen M. (ed.). 1987. *Ivory Scales: Black Australia and the Law*. Kensington: University of NSW Press.

Healy, Bill, Tim Turpin and Margaret Hamilton. 1985. "Aboriginal Drinking: A Case Study in Inequality and Disadvantage," *Australian Journal of Social Issues*. Volume 20, number 3, pp.191-208.

Hennessy, Peter. 1988. "Aboriginal Customary Law and Sentencing," *Aboriginal Law Bulletin*. Number 30. Sydney: Aboriginal Law Center.

ICIHI (Independent Commission on International Humanitarian Issues). 1987. *Indigenous peoples: a global quest for justice*. London: Zed Books.

Kearins, J. February 1991. "Factors Affecting Aboriginal Testimony," *Legal Service Bulletin*. Volume 16, number 1.

LaMacchia, Graeme. 1992. "Mutual Hostility and the Problem of Police Culture in Rural Australia," *Rural Society*. Volume 2, number 4, pp.14-16.

Langton, M. 1988. In Keen, I. (ed). *Being Black: Aboriginal Cultures in Settled Australia*.

Law Reform Commission. 1986. *The Recognition of Aboriginal Customary Laws*. Report Number 31. Canberra: Australian Government Publishing Service.

Lippman, Lorna. 1970. *To Achieve Our Country: Australia and the Aborigines*. Melbourne: Cheshire.

Lyons, G. 1984. "Tactical Response Group," in Brown, D., R. Hogg, R. Phillips, G. Boehringer and G. Zdenkowski (eds.), *Policing: Practices, Strategies, Accountability*. Sydney: Alternative Criminology Journal.

McCorquodale, J. 1987. "Judicial Racism in Australia: Aborigines in civil and criminal cases," in Hazelhurst, Kayleen M. (ed.), *Ivory Scales: Black Australia and the Law*. Kensington: University of NSW Press.

McDonald, G. 1982. *Fighting Communication, Controls and Cognitive Ordering among the Wiradjuri of Inland New South Wales*. Sydney: University of Sydney, Department of Anthropology.

McRae, H., G. Nettheim and L. Beacroft. 1991. *Aboriginal Legal Issues*. Melbourne: The Law Book Company.

Metherell, Mark. 1988. *The Age*, 13 February.

Middleton, H. 1977. "The Aborigines," in Encel, S., A.F. Davies and M. Berry (eds.), *Australian Society*. 3rd edition.

Moss, Irene. April 1991. "The National Inquiry into Racist Violence," *Migration Action*. pp.9-12.

Muirhead, D. 1988. *Royal Commission into Aboriginal Deaths in Custody: Interim Report*. Canberra: Australian Government Publishing Service.

National Aboriginal and Islander Health Organisation. Undated. *Information paper*.

National Inquiry into Racist Violence. 1991. *Racist Violence*. Canberra: Australian government Publishing Service.

Office of the Commissioner for Community Relations. 1981. *Discrimination Against Aborigines and the Role of the Commisioner for Community Relations*. Community Relations Paper. Number 13.

Overview of the Response by Governments to the Royal Commission on Aboriginal Deaths in Custody. 1992. Canberra: Australian Government Publishing Service.

Parbury, Nigel. 1986. *Survival. A History of Aboriginal Life in New South Wales*. Sydney: Ministry of Aboriginal Affairs.

Pittock, Barrie A. 1977. "Towards a Multi-racial Society," in Stevens, F.S. and E.P. Wolfers (eds.), *Racism*. Volume 3. Sydney: Australia and New Zealand Book Company.

Reece, R.H.W. 1979. "The Aborigines in Australian Bibliography," in Moses, J.A. (ed.), *Historical Discipline and Culture in Australia*. Queensland: University of Queensland Press.

Reid, Janice and Deborah Lupton. 1991. "Introduction," in Reid, Janice and Peggy Trompf (eds.), *The Health of Aboriginal Australia*. Sydney: Harcourt Brace Jovanovich, Publishers.

Reynolds, H. 1974. "Progress, Morality and the Dispossession of the Aboriginals," *Meanjin Quarterly*. Volume 33, number 1.

Roberts, Janine. 1978. *From Massacres to Mining: The Colonization of Aboriginal Australia*. London: CIMRA and War on Want.

Rowley, C.D. 1970. *The Destruction of Aboriginal Society: Aboriginal Policy and Practice*. Volume 2. Canberra: A.N.U. Press.

Rowley, C.D. 1971. *The Remote Aborigines: Aboriginal Policy and Practice*. Volume 3. Canberra: A.N.U. Press.

Rowley, C.D. 1986. *Recovery: The Politics of Aboriginal Reform*. Melbourne: Penguin Books.

Royal Commission into Aboriginal Deaths in Custody. 1991. *National Report: Overview and Recommendations by Commissioner Elliot Johnston QC*. Canberra: Australian Government Publishing Service.

Sculthorpe, H. August 1990. "Aboriginal-Police Relations in Tasmania," *Aboriginal Law Bulletin*. Volume 2, number 45.

Searcy, Jennifer. January 1993. *Deaths in Custody Newsletter*. Campaign for Prevention of Custodial Death. PO Box 847, Nedlands, WA 6009, Australia.

Secretariat of National Aboriginal and Islander Child Care. 1986. *Child Abuse and Neglect from an Aboriginal Perspective*. Paper presented to Sixth International Congress of Child Abuse and Neglect, Sydney.

Shallice, Andy and Gordon, Paul. September 1990. *Black People, White Justice? Race and the Criminal Justice System*. London: Runnymede Trust.

South Australian College of Advanced Education. 1985. *Race Relations in Australia*. Adelaide.

Stevens, John. 1987. "Enigma in Black Reform," *The Age*, 17 December.

Stone, Sharman N. 1974. *Aborigines in White Australia*. Melbourne: Heinemann Educational.

Summers, J. 1975. "Aboriginal Policy," in Gibb, D. and A. Hannan (eds.), *Debate and Decision*. Melbourne: Heinemann Educational.

Swanton, Bruce (ed.). 1984. *Aborigines and Criminal Justice*. Canberra: Australian Institute of Criminology.

Tatz, C.M. 1972. "Aborigines: Law and Political Development," in Stevens, F.S. (ed.), *Racism: The Australian Experience*. Volume 2. Australia and New Zealand Book Company.

Tatz, Colin. November 1990. "Aboriginal Violence: A Return to Pessimism," *Australian Journal of Social Issues*. Volume 25, number 4.

The Age. 2 April 1992.

The Age. 11 February 1993. "Cells used for Aborigines inhuman says Amnesty."

Thomson, Neil. 1991. "A review of Aboriginal health status," in Reid, Janice and Peggy Trompf (eds.), *The Health of Aboriginal Australia*. Sydney: Harcourt Brace Jovanovich.

Victorian Aboriginal Legal Service. 1989. "Community Justice Panels: The Koori Initiative," *MOSA Magazine*. Melbourne: Monash Orientation Scheme for Aborigines.

Victorian Aboriginal Task Force on Land Rights and Compensation. Undated. *What All Australians Should Know About Land Rights*.

Victorian Office of Corrections. 1990. *Aboriginal Offenders in Victoria: Recent Statistics*.

Weigel, R.H. and Howes, P.W. 1985. "Conceptions of Racial Prejudice: Symbolic

Racism Reconsidered," *Journal of Social Issues*. Volume 41, number 3.

Wooten, J.H. 1991. "Deaths in Custody," *Current Issues in Criminal Justice*. Volume 2, number 3.

Ageism and Prisonization: Subterranean Forms of Control

Robynne Neugebauer Visano*

> They call you names. You go to the prison store, they snatch the bags from your hands as you leave. You're living in a jungle among savages.
>
> 67-year-old Quentin Brown, Louisiana State Prison
> at Angola (Carroll, 1989: 70).

Introduction: Juxtaposing Exclusionary Practices

The study of prisonization involves the processes of being, becoming and experiencing differences. Prison socialization encourages an appreciation of both the conditions and consequences of being set apart or even relegated to the "margins." Traditionally, the designation of offender refers to a discourse about differences that is used to justify control. Fundamentally, "doing time" in prison is the result of challenging a socially constructed and historically rooted social order.

Clearly, the study of "carceral contexts" is about expressions of power and cultural contests. Penology is about discipline, domination and devaluation. The origin and reproduction of punishment are linked to the social organization of conformity, that is, control. Control constitutes the institutional or patterned responses to differences which are designated as threatening, that is, criminogenic. Society takes crime seriously especially when power or privilege is threatened. Punishment, therefore, emerges as a disciplinary device to control these challenges. For punishment to be effective, it too varies from extremely coercive forms of penal confinement to seemingly more innocuous surveillance strategies. The latter measures are ubiquitous and panoptic — they are the all-scanning and forever vigilant controls that strip the suspect of individuality and create "docile bodies" which are "subjected, used, transformed and improved" (Foucault, 1979: 136) in order to enhance conformity to roles and rules.

Punishment, however, is an extremely elusive concept that defies simple definition. Far too frequently, however, traditional texts tend to conceptualize punishment as congruent with societal expectations. More appropriately, however, punishment articulates trouble, that is, authenticates moral rules and ordered worlds. Normatively, there is a belief held by some that something is wrong in the behavior, views, or even the appearances of others. Therefore, punishment as a felt intrusion, a real or symbolic injury, is legitimated by the requirements that official or formal sanctions respond to the morality of socio-legal demands. But, since sanctions regarding the maintenance of a particular order are always in conflict, punishment remains essentially a judgement grounded in prevailing politics. But, what kinds of punishments are singled out as appropriate and for whom? Definitions of appropriateness tend to focus solely on such normative criteria, standards, or conventions that are limited juridically because they fail to critically address the problematic relationship between penalties and crimes. Again, which crimes matter? And why? Basically, all crimes are created to advance particular perspectives and their meanings are always negotiated among relevant participants/stakeholders. On the one hand, punishment refers to the selection of a negative imputation that stresses the significance of rules and roles which credit certain privileges to this proscription. Punishment occasions the classification of activities, actors and contexts according to guidelines which govern the appropriateness of behavior, attributes, appearances, identities or relations.

Clearly, crime and the concomitant penal responses are cultural products. Culture frames experience, supplies interpretations from which inferences are drawn, legitimates decisions, provides a history and secures a loyalty to rules (Visano, forthcoming). Within the crime calculus the offender is transformed into a marginalized other deserving of exclusion. This construction of the 'other' is determined by dominant discourses — political and legal as well as cultural. Generally, traditional accounts of crime stress the primacy of a binary code; in other words, identities, relations and activities are presented as either deviant or conformist. This artificially bifurcated category misrepresents multi- layered identities or phenomena. Rather, in this article, criminal actors and reactors are equally analyzed as acted subjects and subjected actors, both constituting and constituted within contexts of differential defiance or deference (*ibid.*). Equally, the criminal actor and reactor are active agents situated within wider constituting contests. Actors and reactors, as cultural subjects within discourses of power, are engaged in micro-political (local) struggles shaped by more macro-cultural (global) influences . Criminals are multiple subjects, enjoying or suffering a plurality of meanings that are displaced and re-constructed in concert with other hegemonic reproductions of discipline. Prisonization, therefore, is both an expression of political processes and a consequence of politicized structures. In brief, the study of penology involves the examination of a variety of struggles.

How then do we understand prisonization? The phenomenon of punishment cannot be conceptually appreciated without a fundamental return to the nature of ideas, a concern for knowledge that grounds prevailing perspectives. Such related processes as the understanding and articulation of knowledge tend to govern the contours of penological inquiry while simultaneously shaping commonsensical accounts of punishment. Our task, therefore, requires a scrutiny of what forms of knowledge are advanced as truths by confronting both the authoring subject and the authored text.

Forms of punishment exist within a pre-existing prison subcultures. And yet, the influences of ageism have been overlooked in studies of prison lifestyles. Three related factors are fundamental in building and maintaining the symbolic worlds of prison subcultures: skills of actors, reactions of others and self-identity. According to the prison subculture, survival is conditioned by the acquisition of skills, knowledge and resources. Inmates learn a stock of beliefs, values, and ways of acting that will ensure continued acceptance and participation.

Another important element within the prison subculture is the reaction of significant others. These significant others are instrumental in securing access to resources. More significantly, associations with similarly circumstanced others serve to "validate" (Lemert, 1972) appropriate self concepts. By attending to the reactions of others, inmates learn favorable definitions of experience and self which, in turn, guide new strategies of interaction. Inmates acquire their roles by interpreting the role of others and by reacting to how they think others conceive normal action in a situation. The response of significant others will be incorporated as part of the inmates' collective and individual identities. Another principal influence in advancing one's acceptance in an inmate culture career is the acquisition of techniques for constructing appropriate self concepts. Seasoned inmates establish meaningful identities and self-esteem for themselves. Features from the outside settings that are imported from street values and inherent features of deprivation intrinsic to a totalitarian setting are negotiated within socialization processes that facilitate the reproduction of the necessary world view, skills and knowledge. The logic of this normative proscription presupposes that once actors have begun to move in the direction of a specific cultural mappings they will inevitably survive reasonably well. But some inmates have difficulties dealing with situational, structural and subjective contingencies. A large group of inmates who have hitherto remained ignored in the literature suffer considerable disadvantages because of their age.

Labelling: The Politics of Ageism

In our culture, the conception of seniors is a reflection of socially constructed beliefs and values. Attitudes about aging and seniors tend to be negative (Green, 1981). An ageist framework incorporates attitudes, beliefs and values that seniors are unproductive, incompetent, overly dependent, senile, unattractive and asexual. Moreover ageism, the discrimination based on age, further results in the formation of further negative myths regarding the aging process in general (Friend, 1991).

Ageism has become a contagious phenomenon that erodes a sense of confidence in the older generation. Ignorance about aging has a negative impact on societal members. Ageism has generated counterproductive responses such as alarmingly high levels of anxiety which has led to "a cocooning" orientation — attempting to stay young and reframing oneself according to youthful images.

The labelling perspective is instructive in directing attention away from the causes of ignorance and prejudice, focusing instead on the relationship between people who have the power to label (Schur, 1971, 1980), and those who are unable to challenge the social repercussions of this stigma. Ageist labelling comes to be defined specifically as a process of interaction between at least two kinds of categories based strictly on chronological age: those people or institutions who allegedly commit a "deviant" act or maintain "deviant" attributes, and the rest of society. The two groups are complementary and cannot exist without each other (Becker, 1964). Central to the labelling perspective is a concern with interactive processes of typification and designation (Fleming and Visano, 1983). What becomes central to this process is the reaction of 'some' towards 'others.' As Kitsuse (1962) notes, it is the response to deviance that transforms people. Definitions are reactions to stereotypes that are articulated within the backdrop of a shared universe of symbols.

Typically, socially distanced or marginalized people are treated in a manner that renders them even more vulnerable. Moral entrepreneurs affix stigmatizing labels to those who threaten their normative conceptions. Stigmatization or the collective designation of discredit has far-reaching consequences for an individual's self concept. A label once affixed by powerful interests not only crystallizes identity but is so sticky that it becomes difficult to remove. Specifically, labels generate new definitions of 'self' for the particular person, his or her reference group and the larger societal audiences. Furthermore, if individuals do not attempt to shed the designation, they may 'personalize' or internalize the labels by organizing their lifestyles around the assumptions associated with the labels. A 'master status' emerges and overwhelms all other aspects of identity.

In the context of an interpretive framework, the essence of the ageist label is both definitional and interpretive. These formulations suggest that labelling is not only constituted through an interpretive process but is also a consequence of the application of rules by others. The nature of response is restricted to the application of rules — rules which exist independent of the labeller and independent of the actor who is doing the labelling.

This conceptual perspective moves outside of a static and deterministic approach to rules by examining deviance as constituted by actors in an active and negotiated process. Generally, actors are involved in producing and developing reactions according to an interpretive process of aging. The concept of aging refers to the progression of related experiences and identity changes through which one moves during his or her life stages. Within the normative traditions aging is a socially recognized process involving a relatively orderly or fixed sequence of appropriate age-specific movements. This constellation of behavior and values serves as a framework for interpreting action (Hughes, 1937; Visano, 1987) and for charting identities (Rock, 1979). Within the interpretive paradigm, aging is not only a way of being but also a way of knowing. Traditionally within the dominant social order, aging consists of forms of sociation which impose some intelligibility on the actor's world. Aging focuses on the processes of, and stages in, choice, development and transformation which enable a discernment of age-specific contingencies affecting the nature of interactions and relations. Age-specific stages are characterized by identifiable and organized sets of relations and social meanings. As they emerge from social interactions and as they are subsequently interpreted as meaningful, these chronological anchors provide actors with a set of perspectives. In other words, actors construct knowledge of their different worlds by assessing situations and assigning meanings to the activity and to others in the form of classificatory schemes. Consequently, aging has become an ongoing empirical process of self-indication and self-validation. These categorizations establish routine rules for interaction and serve as directives for future involvements. Regrettably, the above discursive overview of the process of aging is used to define social reality on the basis of a belief in the "objective facticity" of competence and appropriateness. Rules of behavior, codes of conduct, fixed discipline and reward structures become commonsensically accepted as objectively established especially in reference to a body of culturally given norms. The dominant ideology has created automatic, lifeless and generic stages which justify degrees of involvements. These stages as contingencies appear both as conditions and consequences of interactions.

As with all other institutional contexts of modern society, aging also appears to be a horrendous liability in the crime and punishment industry. The value placed on youthfulness within the prison culture shapes the selection of targets and performances. Traditionally, texts on penology have typically overlooked older inmates or simply depicted them as a mass of undifferentiated 'old timers.' In an effort to partially fill this gap and thus provide a more balanced account, this paper explores the implications of aging in penal communities. Specifically, this paper argues that punishment incorporates subterranean forms of control attendant with the prison culture. In addition to official legal sanctions or administrative rules, there are well organized informal, invisible and discretionary 'justice' systems that are not only recognized but encouraged with impunity by prison authorities and inmates alike.

Age, it is submitted, influences the negotiation of roles and rules within prison life. The victimization of older inmates is understandable given the general antipathy towards older inmates among officials and the younger inmate populations. This exploratory article is part of a larger ongoing longitudinal project designed to investigate the phenomenon of aging in reference to a plethora of relations, activities and identity transformations. The difficulties encountered by older inmates are analyzed in terms of wider issues — the reproduction of cultural values such as the glorification of youthful bodies. This current study, however, on seniors and prisonization incorporates a multi-method design of informal interviews with 20 former older offenders and their respective relatives; diaries of five older current inmates; a random sample of 30 younger former inmates; and interviews with health care officials, correctional officers and police officers. The sample consisted of 25 offenders (former and current) over the age of 65 (the "older inmates") and 30 offenders in their 20s (the "younger inmates"). These respective groups identified themselves as older and younger; they routinely employ the above age-specific categorizations.

The Prison Culture: Images of Youthfulness

As a result of repeated interactions, newcomers old and young learn survival strategies from more experienced inmates. These methods include the construction of appropriate typifications; the establishment of secure loyalties; the acquisition of roles, values vocabularies, motives, etc. With the reproduction of inmate codes, congenial perspectives emerge that not only 'make sense' but serve to organize values that are different from and in opposition to prevailing institutional norms. This collective wisdom validates perspectives and authenticates experiences. The prison culture, replete with myths, provides a general frame of reference which is used to justify survival. Moreover, the subterranean values of the prison culture act as a defense mechanism which protects inmates from the negative attitudes of outsiders toward them. But, prison populations are fragmented and differentiated societies; inmates come into contact with many different people and attend to the responses of those who share their values.

A central element of the prison culture concerns youthfulness. In general, the prison culture devalues older inmates. Age is synonymous with the conditions of strength, the physical body. For inmates, a sense of survival recasts the physical body as a resource. Interestingly, one's body is objectified as a weapon. The body is displayed in open shirts, rolled-up sleeves, tattoos adorning their muscles, to name only a few. Within the economy of survival, the body is presented as a weapon to ensure personal safety as well as a resource for procuring benefits. In addition to bodily images, youthfulness as strength is projected in general attitudes towards crime. For the younger inmates, survival in prison is an adventure.

A preoccupation with youthfulness sets limits on the roles of older inmates. Older inmates are dismissed as incapable of meaningful participation within the subculture. They are typically rejected as "losers," "has beens" and "old timers" who symbolize failure.

Younger inmates perceive themselves as tough-minded opportunists who extol the bold virtues of the prison culture. They are more selective in their choice of survival companions. For example, Harry, a 24-year-old former inmate, describes the situation as follows:

> You see this is our shit here. To make it, you gotta have good buddies not some old fart who's pissing around about the old days. He's a fuckin' loser. They can't watch your back. He's gonna have a heart attack or something (July 25, 1993).

This preference for younger and stronger "buddies" signifies an emotional over-involvement with the survival ethos. Likewise, Hank, a 31-year-old former offender, suggests that older inmates are stigmatized because they cannot respond well to the pressures of prison and often retreat from meaningful activities:

> Don't get me wrong, they just don't belong here. A joint's not a old age home. They're cool by themselves; you know, friendly, talk about their kids. But sometimes they're stupid, too stubborn. They gotta keep their mouths shut. Just don't belong. You know, pretty weak.

Time to move over (June 15, 1993).

Despite the wholesale discrimination against older inmates, older inmates themselves remain committed to prevailing cultural values. Given their relatively small numbers, their deteriorating health and their silenced voices, it is not surprising that they are forced to comply with both the institutional rules and subcultural values. Long-termers account for half the population over 55 years of age; the other half committed crimes later in life (Carroll, 1989: 70). Their crimes vary from white collar offenses to acts of direct violence. In this paper older child molesters are not discussed because of the difficulties in securing a sample and because of the general odium directed at them by the inmate population not as a result of age as much as of the nature of their insidious violence.

Older inmates, nevertheless, respond to the pressures of exclusion by adopting techniques that invite more favorable assessments. They move beyond stigma management by protesting and resisting designations of weakness. They continue to socialize with newcomers to whom they fabricate harmless stories about their vigor and past accomplishments. To some inmates they become supportive 'godfather figures.' But their acceptance is mediated by the nature of their crimes and the networks they enjoy with powerful groupings. According to all the inmate interviews, older inmates who are 'connected' to organized crime activities are not harassed because of the respect and protection they enjoy both inside and outside of the institution. That is, they use their age to reinforce a 'wise guy' image to the younger and more naive newcomers. As Stephen, a 66-year-old current inmate suggests:

> I see myself as more mature, more experienced. The jokes don't bother me. I know I'm an old fool. After a little while I stopped hanging around the kids and got involved with the library. Sometimes I help a kid write, set him straight (July 17, 1993).

Likewise, Sam, a 69-year-old former inmate notes:

> Sure you fire back. But it's dangerous. It doesn't take much for some of the younger cons to bash your head in. They enjoy it. They have friends. I come and go. Hang out. I can't compete at their level. But they can't compete at my level either (August 2, 1993).

Sylvester, a 71-year-old former inmate adds:

> The old timers are pretty sick. They complain a lot about arthritis, about this or that. Who's interested in hearing about that? They can't hear, can't see, can't do much. They don't belong there. Everybody knows that. So they stay in the shadows. Hiding and out of the way. Out of the wars and the dealings. It's too dangerous for them even to say something (July 4, 1993).

As the above excerpts highlight, violence is commonplace especially against a recalcitrant weaker inmate. Older inmates are less likely to be violent and jeopardize their parole eligibility status. As Stuart describes: "our days are numbered, we're old and want to be out. There is not much time to lose." Consequently, they can be easily persuaded to defer to the tougher inmates. Their vulnerability places them 'in' the world of prison but not necessarily 'of' it. For older inmates, the younger inmates are irrational, opportunistic, short-sighted prison survivors engaged in dangerous "con games," "hustles" or "scams." Since older inmates are not committed to these role expectations, they are easily subjected to considerable psychological, emotional or physical abuse. They believe that as potential 'targets' they are not well protected in the wider prison community and the administration because they are perceived as double failures. The prison culture permits older inmates to participate in less meaning-ful and harmless prison activities: housekeeping, clerical library work, counsel-ling, assisting in paperwork, correspondence, etc. In turn, the short-term older inmate, who is extremely marginal in the prison community, seeks only fleeting relations with other inmates. With the exception of more long-term violent offenders or those involved with organized crime, older inmates are generally viewed in the culture as docile bodies that can be easily manipulated. In general, there is an apathy towards older inmates that easily turns into contempt whenever the latter express their opinions that openly contradict the accounts of younger inmates.

Interestingly, older clients also debase themselves by assuming a degree of blame. Feelings of shame contribute to their own pacification. They delude themselves into thinking that speaking about their health and age will reduce, if not eliminate, negative stereotypes. Five of the interviewed "older inmates" noted that they would pathologize their situations in order to garner some favorable sentiments. Although they feel rejected, older clients claim that they can "work it out" with younger inmates. Counselling younger inmates has become a safe and emancipatory option that liberates them from discredit. But, once the younger inmates discover that their stories are repetitive, boring and exaggerated, they taunt the older inmates whom they expect to remain passive and deferential despite the insults.

Older inmates tend to accommodate to the pressures in prison. In face of this ageist tyranny of retaliation, verbal abuse and physical threats, resistance is a difficult alternative for older inmates. Facile accommodation also invites offensive or abusive treatment or rejection. Older inmates feel greater insecurity and express serious reservation about their personal safety whenever exchanges occur in more public settings like the workshops, dining halls, shower rooms, etc. In these latter sites, younger inmates seek to "flex their muscles" and demonstrate publicly their power over these inmates. The older inmate is expected to publicly confess his stupidity, negligence or lack of respect. As Foucault (1979) argues, a "confessional" examination is more effective than physical punishment in achieving power and discipline. During interviews all younger inmates candidly admitted taking some advantage of the older inmates and expressed a willingness to use force if necessary to secure compliance.

Aging as a Carceral Context

Both older and younger inmates fully realize that the prison culture values youthfulness. This orientation legitimates the segregation and alienation of older inmates. This celebration of youth is consistent with and derivative of wider cultural values carried to an extreme. Youth-oriented cultures pervade all social relations. Myths flourish as cultural markers which degrade older persons as pathetic. These stereotypes reproduce as well as reinforce ageism. This imposing character of the dominant culture of Western societies is apparent in the discriminatory attitudes of prison subcultures, that is, subcultures that ostensibly claim to be counter-hegemonic.

Prisonization is a site of cultural reproduction. Positive self-identities for older men are difficult to attain within a culture that sanctions appropriate age-specific behavior. Culture enhances ageism, a form of elder abuse. Ageism is integrally related to the process of designating deviance and determining punishment. In fact, the phenomenon of ageism is a resource for formulating and formatting practices of exclusion. Traditions of ageism construct and commoditize discourses that legitimate further stigmatizing behavior. As noted earlier, ageism differentiates, marginalizes and negates identity. Of central concern to our discussion of the fundamental links between ageism and prisonization are the following issues: the culture of ageism; anti-ageist ideologies; the institutional or administrative practices; social organization of custodial cultures; and ageism as a complex of social meanings constantly shaped by both internal (local) processes and external (outside) pushes and pulls which contribute to alienation, powerlessness and exclusion.

Ethnocentrism within the penal system functions to punish, foster solidarity, secure collective conformity and justify further exclusionary exercises. The prevailing ideologies, perceptions and values within facilities incorporate strategies of survival within an assimilationist model. Older inmates are compelled to abandon distinctive age-specific/cultural traits in favor of widely held dominant values. Ageism is an ideology, not simply of differences, physical or cultural, but an ideology of superiority that denies the meaningful participation of older inmates. Ageism, therefore, is not simply an incidental element of the social order within institutions. Rather, ageism goes to the very core of political, cultural and economic conditions.

Summary of Substantive Findings

- Older inmates admit that they are ill-informed about available services.

- Access to information is contingent upon the strength, variety and number of network ties within the prison community.

- A majority of older inmates noted that despite their complaints, the administration fails to acknowledge their concerns.

- Older inmates are discouraged to voice any complaints for fear of retaliation.

- Community resources are not readily available to older inmates, in spite of their growing numbers.

- All older inmates agree that prison staff fail to act according to the role of helping professionals.

- The kinds of health services, social services, retraining etc. delivered are inappropriate for different age-specific groupings.

- All respondents indicate that there are few specific programs/services directed to older inmate groups.

- All older inmates note that there is considerable negligence and ignorance on the part of custodial and administrative personnel.

- All respondents agree that older inmates pose little or no threat to the wider/outside communities and ought to enjoy an early release program.

- All respondents recognize that older inmates often have no families, no jobs and extremely limited options.

- There is a general tendency for younger inmates to suggest that the hostility towards older inmates is an institutional as well as a cultural problem.

- A more holistic, long-term and multi-tiered community response is long overdue, rather than isolated state-identified solutions (identified usually by local prison staff).

- Correctional institutions have not given priority to improving access to their services for older members of the prison population.

- Cultural factors are important for understanding the reluctance of older inmates to seek supportive services.

- Mainstream supportive services for inmates fail to recognize diverse cultural values.

- Very few inmates, old or young, had anything positive to say about both the institutional programs and the staff who served them.

- Inmates believe that the administration does not have a broad enough perspective to respond appropriately to their service needs.

Wider Contexts of Corrections: Generic Implications

As the data from interviews clearly demonstrate, corrections facilities fail to provide a comprehensive social development model. Older inmates (former and current) have generously commented on the hopelessness of their respective life chances once in prison. In addition to the 'passing their time' as unobtrusively as possible, older inmates, are expected to defer to extant practices. Inequalities persist. The interests of correctional service providers and their clients, especially of older inmates are not in sync, despite the protestations of administrators and front-line workers. Perlman (1975) contends that service providers and clients have obviously different orientations. The behavior of providers is mainly circumscribed by the need to be economical in the use of available resources and administrators and policy makers will seek maximum organizational control of resources and programs. Acknowledging that service systems seem to conspire against clients, and that their respective agencies and clients appear to have different interests, it is incumbent upon 'equality seekers' — political activists and prison reformers — to grapple with the issue of access to ensure that all people get the help they need. But as Galper (1975) suggests, services are not designed to promote social change. Indeed, a fundamental function of correctional services is to retard, if not contain changes (McCormick and Visano, 1992). Organizational convenience rather than conscience grounds the efforts of administrators who are overburdened with political interference, law and order campaigns, overcrowded penal systems, underfunding, and a reactionary general public, etc.

This exploratory research investigated the perceptions of inmates in terms of the treatment of older inmates. This study was guided by a community development approach, involving participants in a variety of ways throughout the process of the research. In this social action research, participants were involved in diagnosing problems, collecting information to make necessary recommendations and evaluating the effectiveness of these proposed changes. Participants became advocates for change, that is change agents. From these discussions it was obvious that few institutions/organizations devote much energy to effect change beyond the identification of barriers. While prison authorities express a limited appreciation of the problems inherent in age-specific discrimination, the results are disturbing in terms of actual accomplishments or even the mere adoption of a specific program of action to reduce ageist barriers. Correctional facilities do not effectively respond to ageism. Moreover, institutions operate to promote many solitudes within separate systems, distinguished for example by age, race, sexual orientation and gender, hardly taking into account the diverse needs of their prisoners. Interviews confirmed that a majority of older inmates experience considerable difficulties in adjusting to their respective incarceration. In fact, they are further punished beyond legal proscriptions. They tend to attribute the failings of the penal system to such institutional obstacles as inadequate communication; lack of information; inappropriate management styles and techniques of interaction; lack of understanding of subcultured nuances; complicated hidden economies; the inadequate provision of educational, recreational and job training programs, etc. Older inmates expressed feelings of uncertainty, powerlessness and distance towards a correctional (custodial and counselling) staff. It was also found that subcultural factors either inhibited older inmates from approaching service providers or effectively negated the value of such assistance.

The data suggest that there is widespread discontent among older inmates about fellow inmates, staff and the services obtained. The most frequent criticism directed at staff and fellow inmates concerns their failure to appreciate the predicament experienced by older people. Older inmates suffer the legal penalties imposed as well as the more subterranean forms of punishment inflicted by the inmate subculture. The treatment of older inmates in Canada is historically rooted in ageist or youth-centric assumptions that equate age with competence, prowess and knowledge. Consequently, institutional penal practices and subcultural norms punish older offenders. Hypothetically, even if penal institutions were to be operated by well-intentioned and culturally sensitive administrators, inequalities would be perpetuated because of their reluctance to interfere with routine inmate encounters, that is, the informal prison subculture. Social injustices are experienced in everyday exchanges. Feelings of frustration, alienation and despair are frequently expressed by a powerless older inmate population. But the indifference of governments to the genuine concerns of the excluded seldom results in rage. State initiatives not only continue to be insensitive to these differences, but more significantly, they are designed to further criminalize the already severely marginalized. Based on the above policy implications, the following recommendations at a minimum warrant acceptance, given the cruel and unusual punishment suffered by an aging population of inmates.

Recommendations: Institutional Remedies

- There is an immediate need for increased public consciousness regarding ageism in prisons. Currently, there are no collaborative practices let alone a shared vision to reflect equal and responsible entitlements.

- There is a need for older inmates to be politically organized in order to articulate their own needs and resources for health services, parole eligibility, community support, release programs and welfare provisions.

- Provincial and federal authorities need to be committed to an anti-ageist agenda with policies that are clear, consistent and reflective of the input from various constituencies and community-based organizations.

- There is a need to establish more coherent and multi-tiered strategies designed to alter current carceral structures in order to organize and maintain structures of access.

- Although the Charter rhetoric is overwhelming in describing the entitlements of all citizens to public services and resources such as health, there are numerous collectivities and groupings which are deprived of fundamental entitlements. There is a need for a Charter challenge on behalf of impoverished groups.

- There is a lack of commitment on the part of state agencies to confront the realities of exclusions.

- There is a need to make organizations and public institutions that provide rehabilitative services more reflective of Canada's diversity.

- There is a need to develop a demographic profile and follow-up of older inmates as well as a clearing house of services, resources and self-help groups to be established for older inmates.

- There is a need to eliminate specific barriers to equity and access in existing state organizations that are ageist, sexist, homophobic, racist and ableist.

- There is a need on the part of community-based organizations to develop a critical agenda that departs from the rhetoric and myths of official legal practices. Central to this orientation is an interrogation of current configurations of power, the political economy, the ideology of capital/consumerism and the culture of controls.

Endnotes

* The author gratefully acknowledges the extremely generous advice, support and responses provided by the many anonymous participants. A special note of gratitude is extended to R. Lemick, L. Reti, R.Doyle, the Vita Nova Foundation and Club Avanti.

References

Alhonte, M. 1981. "Confronting Ageism," in Tsang, D. (ed.), *The Age Taboo*. Boston: Alyson.

Becker, H. 1964. *Outsiders: Studies in the Sociology of Deviance*. New York: Free Press.

Carroll, G. 1989. "Growing Old Behind Bars." *Newsweek* (November 21):70

Fleming, T and L.A. Visano. (eds.). 1983. *Deviant Designations*. Toronto: Butterworth.

Foucault, M. 1979. *Discipline and Punish*. N.Y.: Pantheon

Foucault, M. 1980. *History of Sexuality. V.I.* New York: Vintage.

Friend, R.A. 1991. "Older Lesbian and Gay People: A Theory of Successful Aging," in Lee, J.A. (ed.), *Gay Midlife and Maturity*. NY: Harrington Park Press.

Galper, J. 1975. *The Politics of Social Services*. Englewood Cliffs: Prentice- Hall.

Green, S.K. 1981. "Attitudes and Perceptons about the Elderly: Current and Future Perceptions." *Aging and Human Development*. 13: 95-115.

Hughes, E. November 1937. "Institutional Office and the Person," *American Journal of Sociology*. 43: 409-410.

Kitsuse, J. 1962. "Societal Reaction to Deviant Behavior," *Social Problems*. 9 (Winter): 247-256.

Lemert, E. 1972. *Human Deviance, Social Problems and Social Control*. Englewood Cliffs: Prentice-Hall.

McCormick, K.R.E. and L. Visano. 1992. "Corrections and Community (In)action," in McCormick, K.R.E. and L. Visano (eds.), *Canadian Penology: Advanced Perspectives and Research*. Toronto: Canadian Scholars' Press Inc.

McPhersen, B. 1983. *Aging As A Social Force*. Toronto: Butterworths.

Perlman, R. 1975. *Consumers and Social Services*. New York: J. Wiley and Sons.

Rock, P. 1979. *The Making of Symbolic Interactionism*. London: Macmillan.

Schur, E. 1971. *Labelling Deviant Behavior*. New York: Harper and Row.

Schur, E. 1980. *The Politics of Deviance* Englewood Cliffs: Prentice-Hall.

Visano, L. 1987. *This Idle Trade*. Concord: Vita Sana Books.

Visano, L. Forthcoming. *Beyond the Text*. Toronto: HBJ- Holt.

Prisoners of Their Own Device:
Computer Applications in the Canadian Correctional System

Kevin R.E. McCormick

Disciplinary power is exercised through its invisibility; at the same time it imposes upon those whom it subjects a principle of visibility. In discipline, it is the subjects who have to be seen. Their visibility assures the hold of the power that is exercised over them. It is the fact of being constantly seen, that maintains the disciplined individual in his subjection (Foucault, 1986: 187)

Introduction: Electrifying the Carceral

Currently, the Canadian penal system is being shaped by an all-consuming infatuation with the emancipatory possibilities inherent in each new technological innovation. The introduction of devices such as electronic inmate monitors and offender management computer systems designate linguistic, mechanical and conceptual contours, electronically translating traditional perspectives of incarceration and rehabilitation. Computer applications in carceral contexts electrify the correctional process and in doing so technologically translate the penal agenda. The purpose of this paper is to facilitate a critical comprehension of the role of electronic technology within various facets of the correctional system, including prisons, parole and probation and rehabilitative contexts. This penal-technological investigation will detail clearly the extent to which the correctional process has become electrified and situate this discussion within issues of inmate care, staff development, institutional management and community safety. It will be demonstrated that technology is an exercise in control despite the universal celebration of its liberating potential and the participation of a consensually oriented technocratic society. As a cultural marker, technology mirrors a generic deference to authority, electronically translating care/control practices, management principles and community interests within Canadian society.

Technological Applications and Carceral Contexts

The social introduction and technical application of computers into the formal sphere of social control has translated technology into a formidable resource in the arsenal of state coercive exercises (Ackroyd *et al.*, 1977). In all facets of society, both formal and informal, control mechanisms have been expanded as a result of various technological innovations. Heise (1981) forecast that technological implementation would dramatically impact upon the very nature of social existence:

> by the end of the decade, microcomputers will have changed the way social scientists do research, the way they teach courses and the way they work in applied settings. And computers will also create new topics for social analysis as the microcomputer revolution reaches diverse sectors of society (1981: 395).

Physical computer systems and their attendant operating systems have become the center point of most social interactions in society, ranging from leisure, medicine, education and the criminal justice system. Technological innovations designate for the users lexicons and electrified norms of etiquette, electrified parameters in which the user must develop and execute strategies of social interaction and articulation. As a result of this imposed 'techno-contract' between the apparatus and the user, a disproportionate power dynamic is initiated. Both the experiences and methods of social presentation utilized by the user are imprisoned in a technologically mediated environment. While the user felt that this agreement was a small expense considering the benefits promised by the device, the influence must not be merely considered causally utilitarian, but observed that "repressive technologies were developed to help suppress the revolt of a partially subjugated caste" (Ackroyd, *et al.*, 1977).

In submitting to these lexicons, conceptions and definitions, human experiences are organized according to the state agenda and translated through a socially articulated and highly coercive technological script (Pfaffenberger, 1988; Meyrowitz, 1985). This agenda is not readily manipulated by the user, but rather causes users to articulate themselves socially and exclusively through the techniques prescribed by the apparatus. Users of any technology must avoid becoming hypnotized by the convenience afforded them by the device and realize that the relationship between the user and apparatus is not utilitarian, but rather insidious in its carceral impact. Emancipatory assurances of the computer in correctional settings are hollow cliches, couched in the rhetoric of convenience and firmly entrenched in the inequality of the political.

The Convenience Myth: Electronically Erected
Stages of Social Performance

In general, technology is manifested according to functions which facilitate an economy of convenience in the performance of tasks. Regrettably, the latent insidious repercussions attendant with the social introduction of technology, like the imprisonment of the user within imposed definitions of the specific situation, are far too frequently overlooked. This techno-carceral process limits social interaction "configuring the user — defining the identity of putative users and setting constraints upon their likely future actions" (Woolgar, 1991: 58). Thus, the relationship between technology and the user is inherently carceral given the ubiquitous nature of conflict within power imbalances. Asymmetry establishes dynamic stages of social interaction upon which a myriad of "technological dramas" (Manning, 1992; Meyrowitz, 1985) are enacted routinely.

Beyond mere enhanced levels of convenience in the control industry, the computer is transforming the nature of social interaction. This specific technology used by state functionaries is not a separate entity from the individual, used and then placed down, but rather in its utilization becomes part of the person, "a dynamic by which the technology becomes an extension of the user's central nervous system" (McLuhan, 1964: 64). By extending the scope of their senses, the proclivity to perceive a greater spectrum of opportunities also increases. These sites of interaction are not physical settings, but constitute "virtual communities," cultures comprised of electronic impulses housed in technical casings. Agents of formal control present themselves electronically through the medium of the computer; they perform before an electronically constructed and politically constituted audience with well-defined organizational expectations. What is generated is not merely a re-articulation of techniques of performance used when depicting self in an intimate social circumstance, but rather a societal presentation contained within and constrained by the specific vocabularies, protocols and precepts attendant with the technology.

While extant technological power differentials are evident in all aspects of life, the negative impact is more fully appreciated whenever the device fails to emancipate and succeeds a method of social and technological incarceration. The processes by which individuals first become imprisoned in "technological barriers" and subsequently internalize the apparatus's unique lexicons, as methods of social articulation and definition, are not temporally or spatially restricted. Conceptually, control is not situationally-specific, but rather a social context that characterizes the relationship of the device and the user. While assisting the user in certain tasks, technologies readily transcend the "convenience function" by imposing on the user an electronically-defined and mediated framework that comprehends and traverses immediate social environments presented by technology. Traditionally, users of devices have disregarded the carceral implications inherent in technologicalization. Rationalizing technology as a mere cost-saving program blatantly ignores "technology as a system of exploitation" (Buchanan, 1965: 535). In order to be "convenienced" by the device, the user suspends and brackets his/her personal experiences and modes of social presentation. The seductive powers of the technology unconditionally submit the unsuspecting and at times euphoric user to a foreign dogma replete with rituals, a catechism of lexicons and a set of reconstructed societal values and beliefs (McCormick and Visano, 1992).

Hysteria and hypnotic effects have accompanied the social introduction and implementation of every technical device. For example, the computer continues to be heralded in the popular culture as the "greatest invention since the wheel." This infatuation with the role of computers in enhancing everyday lives tends to overshadow current theoretical debates regarding the appropriate role of this device within the wider social order. Levin (1986: 116) argues for a critical appreciation of the technology's introduction, noting that "social scientists still wonder whether the computer, rather than people, has dictated the trajectory of sociological theory in recent decades." Instead of blindly embracing the ease by which certain mundane tasks can now be accomplished by the computer, social analyses of this technology must not be fixated only on the "liberation" thesis. A critical evaluation of the ways in which the computer captures the user in a prison defined and mediated by "the technological order" (Ellul, 1976) is warranted.

Techno-Control: Computers as Formal Control Agents

The carceral aspect of technological implementation is most graphically illustrated when examining the introduction of electronic devices into the criminal justice system, (Manning, 1992; Colton, 1979) specifically in the field of penology. As a direct result of computer implementation, the penal process has ushered in a new dimension of social control in which expectational structures and performance techniques become technologically defined and articulated. This technological revolution was celebrated within the liberal rubric of convenience, subscribing to a humanist techno-orientation. The time saved on paperwork will result in greater time spent on the direct service side of the penal industry (Campbell, 1989; Gooderham, 1986). Within the physical prison the individual's movements are observed within the parameters established by the prevailing political and economic sentiments of state. Introduction of electronic devices into the formal process of punishment translates directly the physical manipulation and isolation of inmates into a technological sphere. For years, "electronic surveillance" (Whitehead, 1992; Clear and Hardyman, 1990) had been celebrated within "modern penology" as contributing immeasurably to cost saving and efficiency-enhancing initiatives. Interestingly, these new devices also act as constant reminders to the offender. Given that they are physically attached, they act as an ongoing visible reminder to the offender of their responsibility to an omnipresent process of social control. These apparatuses of penal servitude are technologically restricting and physically oppressive. We are witnessing the technological translation of carceral conditions in the physical institution to the widespread generation of "electronic jails" (Berry, 1985).

Technological implementation within the Canadian penal system seemingly "liberates" authorities as they welcome a commitment to convenience. For the (x)offender, technology serves to render even more invisible the following: rules, pervasive conditions and overall patterns of accountability of anonymous state officials. On the other hand, the authority of policing agencies is calibrated according to resources (Manning, 1992; Colton, 1979; Colton and Herbert, 1979). Technology as an empowering resource structures discretionary practices. In turn, those within the carceral culture are subordinated by the normative contours of the techno-culture, cast into an environment of their design, but beyond their control. This process of techno-incarceration is witnessed in every facet of the Canadian penal structure, including prisonization, decarceration, rehabilitation and education programs for both inmates and care/control agents. Any sacrifice offered to this electronic guard by those seeking electronic liberation only further incarcerates the user within a technologically invisible panopticon of impulses far more carceral than any physical mode of surveillance and control.

The Electronic Cell: Penological Implications of Technological Innovation

The introduction of electronic devices such as computers into the penal system has erected an invisible electronic cell of imprisonment which constricts the actions of both the offender and those agents charged with their monitoring. Rather than serving as a visible reminder to the offender of their relationship to the penal system, the computer's implementation into the process of parole and probation has established an invisible electronic system of technological "decarceration" (Chan and Ericson, 1981; Scull, 1977) where the all-seeing eye of the state becomes directly internalized by the offender and consequently present in all aspects of their existence. Interestingly the carceral context which the computer constructs is systemically overlooked, fixating rather upon the "liberating" capabilities attendant with the apparatus:

> there is a general tendency for scholars and researchers, to ignore or even deny the effects of invisible environments and the latent effects of technologies, simply because they are invisible (Meyrowitz, 1985: 20).

Computerization entraps the (x)inmate within a electronic cage of social articulation, a stage of presentation which, ambiguous in intent, causes the offender to rely upon formal organizations (Clegg and Wilson, 1991) to define the social parameters of their social interactions. The role confusion established by the disproportionate power relationship incarcerates all social performances within an electronic file, not merely the actions which transgressed from the specific conditions of release, but every aspect of their human existence. Infusing itself directly into all actions of the inmate, the computer re-articulates the carceral experiences. Clearly, the constitution of an electronic control perpetuates the cultural values of penal institutions, including compliance with authority and a liberal sprinkling of participation in activities exposing the accredited values of the system. The computer has electronically extended the traditional model of the "panopticon." The daily lives of offenders both within and outside of the institution are under the constant scrutiny of penal agendas, dictated and enforced by an all encompassing "technological eye."

Panopticon of Impulses: Electrifying the Carceral

Far more detrimental in its impact on the entire penal system is the introduction of electronic devices into the very nature of correctional control/ care. The transformation of the carceral culture into the computer age constricts the actions of both the offender and their agents within electronically scripted and politically contextualized agendas. The computer readily transcends the physical reminder of penal servitude afforded by electronic monitoring devices attached to the body of the offender. Rather the individual's body becomes extended electronically through interaction with the computer, erecting a technological stage of performance upon which new forms of penal negotiation are constituted.

Implementation of computers in the penal system imprisons the (x)offender in an technological prison, electronically extending the all-seeing eye of the criminal justice system through the omniscient power of various computer networks, including Offender Management Systems. Imposing itself directly in the daily life of the offender, the computer serves as a constant reminder of carceral servitude. Obedience and complete subjugation to the rules and agents of the prison structure demanded within the confines of the physical institution are electronically replicated by the computer. This electrification of the formal correctional structure technologically extends conditions and practices consistent with the carceral culture. As a repercussion of the computer's social introduction and technical implementation into the penal system, the daily routine lives of offenders both within and outside of the institution become captured, adjudicated and sanctioned by the omniscient power afforded the structure by the technologically magnified panoptic eye of the computer.

Computers and the Correctional System: Electrifying the Penal Agenda

While both the service and policy sides of the penal system have celebrated the benefits of various computer devices and software programs, interestingly, few scholars have paused from the phrenetic electronic pace to reflect on the long-term effects which the introduction of these apparatuses will have upon the intricate process of correctional care/control. While the computer and various software packages may liberate the individual from certain tasks, the more insidious effects of its technological intrusion on the overall correctional process must not be disregarded. What is required is a critical examination of computers and the entire penal enterprise noting the impact that the device has upon all institutional structures (Meyrowitz, 1985; Ackroyd *et al.*, 1977), including care/control officers and (x)offenders (Gomme, 1992; Whitehead, 1992). It is this call for a critical examination which comes out of the hypnotic and game-like effect that computers have had on the penal system.

The correctional system itself has adopted the fundamental precepts of computer emancipation and in doing so has ushered in a new and far more panoptic system of control and coercion. Further, both the physical devices and associated software programs are implemented into the very core of the correctional system as management principles, systematically ignoring the impacts that these devices will have on the lives of all those within the carceral culture. During a 1982 conference of the Canadian Correctional Service (C.S.C.) entitled *Control — Neither Too Much Nor Too Little*, a senior official noted:

> Wardens and other managers were mainly concerned with their number one mandate — the control of inmates. Today that's no

longer sufficient. C.S.C.'s managers are facing a challenging new world where a variety of management controls are essential to administer budgets and staff, as well as inmates (C.S.C. Publication, 1982).

Attitudes such as these raise crucial questions regarding the ethical ramifications of technological applications in the penal system. The introduction of electronic devices into practices associated with community corrections have radically impacted on the very nature of the personal relationship between care/control agents, clients and the community. Whitehead notes:

> The contemporary community corrections worker may be conducting curfew checks or collecting urine samples, whereas yesterday's officer was leading group counselling sessions or referring a troubled probationer to a community counselling agency. Today's officer possibly thinks of the job of community corrections as a job of surveillance whereas yesterday's officer often saw the job as one of service (1992: 155; Studt, 1973).

All sectors of the correctional system, including prisons, detention centers, parole and probation programs have fully embraced the merger of the computer and effective penal management objectives, urging its employees to learn the ease at which impulses may be manipulated. The computer prioritizes the actions of control agents, organizing the penal enterprise according to the technical abilities of the device as contextualized within specific political agendas. The success of computer implementation is premised upon the ability of those designing the systems to distance the user from the personal ramifications of each program executed.

Originally when computers were first implemented as a directly accessible apparatus in the penal system, managers and wardens were lured to the technology with carnival-like promises. A senior correctional official

> urged wardens, parole directors and superintendents attending the National Administrator's Conference to zap space invaders, play electronic blackjack and other computer games to get the feel and scope of the computer terminal (C.S.C. Publication, 1982).

While educational practices would dictate establishing a level of comfort with the computer, this process must not occur to the exclusion of those most dramatically affected by the device, namely inmates and other intimates in the process. Correctional attitudes such as those mentioned fail to adopt a critical appreciation of the technological innovation, which notes that

> new tools and the skills to use them raise significant methodological
> questions: as much as they encourage new kinds of questions and new
> answers to old ones, the real challenge will be to shape the technology
> and use it well (Gerson, 1987: 407).

Clearly, the liberation thesis inherent in correctional computing demonstrates a blind technological infatuation which requires submission to the device's language and definition in order to reap its electronic benefits. What is required is that those charged with care/control responsibilities remain vigilant, consciously attributing a life behind each touch of the computer keyboard. While the new computing care/control officer may feel detached from the process when using the technology, he/she must never lose sight of the effects of the electrified carceral process on the future of those (x)offenders not as technologically empowered.

Computers and the Carceral Context: Electronic Reflections

In conclusion, this inquiry has illustrated technology as a carceral culture and incarceration as increasingly technological. Technology shapes and is shaped by the nature of the cultural definitions. The subject of technology is a problematic discourse that defies simplistic interpretations. It was argued that the convenience of technology reproduces compliance. Seasoned users or technocrats have become the new philosophers/theologians concerned with the hegemonic moralities of convenience. But the phenomenon of technology is about exclusion and resistance, not just accommodations contextualized within the intersections of culture and convenience. Technology is an extension of traditional modes of social control, designed to mirror a generic deference to authority. The dominant order with its emphasis on convenience scripts technology as a commodity that is marketable and profitable for those who have a stake in the penal structure. What warrants further investigation is the extent to which technology itself is contextualized culturally, mediated politically and articulated economically. The user must challenge the myths, monologues and recipes that mediate knowledge and encourage electronic colonialism. Users are not well-informed citizens; knowledge is not available but filtered through self-serving organizations demanding deference to technology. A carnival of illusions and images render the user a vulnerable target of manipulation. Lamentably, the user readily consents to his/her own incarceration and in turn reproduces forms of incarceration in encounters with less privileged and technologically empowered populations.

References

Ackroyd, C., K. Margolis, J. Rosenhead and T. Shallice. 1977. *The Technology of Political Control*. Britain: Pluto Press.

Berry, B. 1985. "Electronic Jails: A New Criminal Justice Concern," *Justice Quarterly*. 2: pp. 1-24.

Buchanan, R.A. 1965. *Technology and Social Progress*. Oxford: Pergamon.

Campbell, Colin. March 1989. "Laptop Computers For Parole Officers," *Let's Talk*. Vol. 14, No. 2.

Chan, J.B.L. and R.V. Ericson. 1981. *Decarceration and the Economy of Penal Reform*. Toronto: Center of Criminology, University of Toronto.

Clear, T. and L. Hardyman. January 1990. "The New Intensive Supervision Movement," *Crime and Delinquency*. 36: pp. 42-60.

Clegg, Stewart and Fiona Wilson. 1991. "Power, Technology and Flexibility in Organizations," in Law, John (ed.), *A Sociology of Monsters*. London: Routledge.

Colton, Kenneth W. 1979. *Police Computer Technology*. Lexington: D.C. Heath.

Colton, Kenneth and Stephen Herbert. 1979. "Police Use and Acceptance of Advanced Development Techniques: Findings from Three Case Studies," in Colton, Kenneth W. (ed.), *Police Computer Technology*. Lexington: D.C. Heath.

Correctional Service Canada. June 1982. "C.S.C.'s Computer Revolution: Greater Management Controls The Watchword For 1982 Says Commissioner," *Let's Talk*. Vol. 7, No. 2.

Ellul, J. 1976. "The Technological Order," in *Technology as a Social and Political Phenomenon*. John Wiley and Sons, Inc.

Foucault, M. 1986. *Power/Knowledge*. New York: Pantheon.

Gerson, Elihu M. 1987. "Do We Sincerely Want To Be Programers?" *Qualitative Sociology*. Vol. 10, No. 4. Human Sciences Press.

Gomme, Ian M. 1992. "From Big House to Big Brother: Confinement in the Future," in McCormick, Kevin R.E. and Livy Visano (eds.), *Canadian Penology: Advanced Perspectives and Research*. Toronto: Canadian Scholars' Press Inc.

Gooderham, Helen. November 1986. "Staff Microcomputers Revolutionize CSC's Way of Handling Information," *Let's Talk*. Vol. 11, No. 13.

Heise, David R. 1981. "Microcomputers And Social Research," *Sociological Methods and Research*. Vol. 9, pp. 395-536.

Levin, M.L. 1986. "Technological Determinism In Social Data Analysis," *Computers and Social Sciences.* Vol. 2.

Manning, P.K. August 1992. "Technological Dramas and the Police: Statement and Counterstatement in Organizational Analysis," *Criminology.* Vol. 30, No. 3.

McCormick, Kevin R.E. and Livy Visano (eds.). 1992. "Technology and Control: Generic Trends" in *Canadian Penology: Advanced Perspectives and Research Toronto.* Canadian Scholar's Press Inc.

McLuhan, Marshall. 1967. *The Medium is the Massage.* New York: Random House.

McLuhan, Marshall. 1964. *Understanding Media: The Extensions Of Man.* New York: Mentor Book.

Meyrowitz, Joshua. 1988. *No Sense of Place.* New York: Oxford University Press.

Pfaffenberger, Bryan. 1988. *Microcomputer Applications In Qualitative Research.* Sage Publications Inc.

Scull, A. 1977. *Decarceration.* Englewood Cliffs: Prentice-Hall.

Studt, E. 1973. *Surveillance and Service In Parole.* Washington: U.S. Department Of Justice.

Whitehead, John. 1992. "Control And The Use of Technology in Community Supervision," in Benekos, P. and A. Merlo (eds.), *Corrections: Dilemmas And Directions.* Anderson Publishing Co.

Woolgar, Steve. 1991. "Configuring the User: The Case of Usability Trials," in Law, John (ed.), *A Sociology of Monsters.* London: Routledge.

Prisoners' Families or Prisoner Families?
A Literature Review

Kevin J. Baker[*]

Introduction

Even under the best of circumstances identifying the socio-economic impact of imprisonment is a daunting task. It can be argued that imprisonment directly or indirectly impacts on everyone in a given society. Of course, the overall effect of the impact is highly dependent on one's attachment to the individual being imprisoned. It can also be argued that identifying (or at least attempting to identify) the impact of imprisonment can only be realistically achieved up to a certain level of remoteness from the individual being imprisoned. No matter where one chooses to draw that line, there is little dispute that there is a need to broaden our understanding of the impact of imprisonment. For prison administrators, policy-makers, and researchers a comprehensive understanding is fundamental and crucial to effective programming and reform. The real challenge is how to make the impact analysis as inclusive as possible, while confronting the very difficult problems created by limited financial and time constraints.

Pragmatic considerations, however, have an unfortunate side-effect on decision-makers. They force administrators, policy-makers and researchers to engage in a process of differentiation; to prioritize who can and cannot be (and/ or who should and should not be) included in an impact evaluation. In turn, this differentiation process promotes a natural gravitation of the researchers' attention and empathy toward those who are the most directly impacted by imprisonment. The pattern is that first priority is given to prisoners. All others who are less directly affected by imprisonment are afforded lower status (Homer, 1979). Over a period of time, the prioritized selection of prisoners as research subjects (at the expense of all others affected by imprisonment) creates an unrepresentative and very narrow understanding of the overall impact of imprisonment. It tends to keep the attention of those who are concerned about prisons focused on conditions on the "inside." The impact of imprisonment on prisoners' families is a perfect example of a significantly impacted group that has suffered because of this differentiation process and the ensuing struggle for the prison researchers' attention (Bakker *et al.*, 1978).

Prisoners' families have been the subject of social-scientific research for years.** However, it was not until the early 1970s that they became the topic of any sustained academic investigation (See Table 2). Since 1970 there has been a dramatic increase in the number of people actually engaged in prisoners' families research. The diversity of their philosophical and methodological approaches have significantly expanded both the quantity and quality of the available literature on prisoners' families. For reasons already noted, however, there remains a significant gap in our understanding of prisoners' families and of *their* issues and concerns. This is especially true in Canada where much of our information on prisoners' families continues to be derived from American and British studies as indicated in the following table (See discussion in Mawby, 1982, pp. 24-25).

Table 1. Country of Prison(s) Being Examined in Reviewed Articles

Country	No.	Country	No.
United States	112	England	22
Israel	4	Canada	3
Scotland	3	Australia	3
Poland	1	India	1
Netherlands	1	Germany	1
Guatemala	1	Multi-national	2

It should also be noted that the term 'prisoners' families literature' is in and of itself deceiving. As you will soon see, a significant quantity of the literature on prisoners' families is not really about prisoners' families but about prisoners with families. This is not to suggest, in any way, that prisoner research and literature is valueless or unjustified. It is simply to reaffirm my initial argument that prisoners' families are being significantly affected by the narrow focus of prison research in general.

One of the primary objectives of this literature review was, quite simply, to identify the major themes or paradigms in prisoners' families research. Another important objective was the identification of some of the more obvious (and perhaps not so obvious) aspects of prisoners' families that were (and are) being seriously under-researched. The ultimate objective was, of course, to provide those interested in this particular area of research with an overview of the work that has been and is currently being done and to reinforce the idea that there is a very real need for their interest and scholarly contributions. In any event, I hope that this article will provide the reader with some insight, however small that might be, into the lives of prisoners' families. The underlying argument being made in this article is that the efficacy of prison and prison-related policies and programs are (or at least should be) largely determined by a comprehensive understanding of *all* the relevant issues.

It is this writer's position that prisoner issues cannot be responsibly determined without any consideration of how prison and prison-related policies and programs affect the prisoners' families. Family relationships are a deeply ingrained aspect of many, if not all, societies including that of Canada. The unique and special status of a family relationship is recognized by and entrenched in Canadian law. Both private and public law, impose special responsibilities, obligations, trusts, benefits, privileges and so on, on individuals solely on the basis of their family relationship to one another. Even when a family relationship is not positive it still maintains a status that is different from any other type of relationship in our society. Under the circumstances it would seem reasonable to conclude that whatever happens to one family member, even if that family member is in prison, directly affects the lives of each of the other family members.

Policy-makers must recognize that serious deficiencies in the existing knowledge on prisoners' families mean that there are also serious deficiencies in their understanding of prison as an institution in our society. Decision-makers must also recognize that very important policy and program decisions, decisions that directly affect thousands of people, are being made with very little factual basis and more often than not on erroneous assumptions. If the goals of prison administrators, policy-makers and researchers truly are to initiate meaningful prison reform and effective prison programs then they must demand *factual* information about *all* of the people that these policies and programs affect.

A number of personal concerns have also been identified and addressed in this paper. A new approach to the study of prisoners' families (including the greater inclusion of alternative or non-traditional families, parents and siblings, and in-laws) is proposed. The commencement of a comprehensive demographic profile of all those being directly impacted by the criminal justice system is urged. And finally, some questions are raised about the potential dangers inherent in the current practice of assigning institutionally-functional values to prisoners' families in order to further their cause.

Methodology

This literature review began with the location of approximately 60-70 journal articles on, to one degree or another, prisoners' families. Using the bibliographic references found in that initial set of articles, another 50 journal publications were identified which, in turn, led to the discovery of approximately another 50 journal publications on the subject of (or related to) prisoners' families. During this process an additional 300 publications, books, dissertations and pamphlets were also identified (Baker and McCormick, 1993).

Due to the volume of literature and the difficulties involved in locating and obtaining some of these materials the decision was made to limit this review exclusively to the "academic" journal articles. The overall number of articles reviewed (combined with the fact that many of the journal articles under review contain summaries of excluded literature and/or are written by authors of the excluded literature) suggests that a fairly representative sample of the overall literature was obtained. In total 157 journal articles were reviewed for this paper.

The selected articles were initially organized according to their particular orientation towards prisoners' families. In many articles, families are treated and promoted as tools available to prison administrators and to society as a whole. It is the potential institutional benefit and value of a family's relationship with a prisoner that is of paramount concern for the researcher, not the family itself. Because the ultimate objective of sound prisoner-family relations is often prisoner conformity and/or rehabilitation, these articles were and are referred to as "prisoner-oriented" articles. The second set of articles attempt examine the social and economic consequences of imprisonment as they apply specifically to prisoners' families. In these articles the prisoner is simply seen as the link between family and the prison. These articles were and are, therefore, referred to as "family-oriented" articles.

Due to the wide range of topics (particularly in the prisoner-oriented articles) the articles were further divided into five categories based on the specific area of emphasis (See Table 2, below). The five categories include Family Life and

Juvenile/Criminal Behavior; The Effect of the Family on the Prisoner's Institutional Behavior; Prisoner-Family Programs; and The Role of Prisoners' Families in Rehabilitation, Social Reintegration and Recidivism; and the Impact of Imprisonment on the Prisoners' Family. As expected, some of the articles defy simplistic categorization. In many instances the difficulty arises because the articles share similar theoretical foundations and approaches. These theoretical relationships are, in fact, very important and will be discussed at length later in this article.

Table 2. Decade of Publication and Major Themes of the Reviewed Literature

Period	Family Life and Criminal Behavior	Family's Role in Prison Behavior	The Family's Role in Rahab.	Prisoner-Family Programs	The Effect of Prison on the Family	Total
Pre-1960 No. of Articles	1	2	1	1	0	5
1961-1970 No. of Articles	2	3	2	1	6	14
1971-1980 No. of Articles	11	7	24	13	17	72
1981-Present No. of Articles	3	9	12	15	27	66
Total	17	21	39	30	50	157

The Literature

Family Life and Delinquent/Criminal Behavior

The literature in this area is one of the most researched and traditional of the various family-criminal justice system topics. While a small percentage of the articles explore the early childhood experiences of adult offenders, most of the literature concentrates on the topic of juvenile delinquency. "Family," in these studies, is used as a variable in the search for a positivistic, cause-and-effect model of criminal/delinquent behavior. Much of the research is grounded in a social psychological perspective and uses a quantitative methodological approach. Generally, the question being explored is: "What aspects of the prisoner's early family life caused or contributed to the prisoner's criminal behavior?" (See for example McCord, 1979; Davies and Sinclair, 1971).

Social factors like early family experience can, and have been, positively correlated with delinquent or criminal behavior in a number of studies (Ousten, 1984; Page *et al.*, 1980; Austen, 1978). It should be noted, however, that the relationships in these studies are correlational and could be the result of any number of related variables (Wilson, 1980). It is not, however, correlational links between early family life and deviance that are of concern at this point.

For obvious reasons, a significant portion of this literature is concentrated on the pre-conviction period of the prisoner's life and does not, therefore, directly address the issues of prisons or imprisonment. It has been included in this review because the "family life-deviance" model of research is extremely important to our understanding of the tradition behind existing theoretical approaches to the study of prisoners' families. There is a strong and clearly identifiable philosophical similarity between the "family life-deviance" literature and the literature that addresses the family's role in the behavior of the prisoner and on the family's role in prisoner rehabilitation and reintegration programs (which form two significant categories of literature on the topic of prisoners' families).

The Family's Effect on the Prisoner's Institutional Behavior

As noted above, this area of prisoner-family literature has strong philosophical links with the "family life-deviance" category of literature. Unlike its forerunner, however, this area of literature does not focus exclusively on the negative effect of the family on the individual. The area has a varied and dichotomous gathering of contributors. On the one side there are the prison administrators and bureaucrats who are concerned about the negative impact that troubles on the "outside" can have on the "inside" (Sterling and Harty, 1972, p. 32). The prevailing view seems to be that the dysfunctional family was a major contributor to the initial deviance and that the quality of these family ties will not improve after imprisonment. It is argued the family will only serve to mitigate the effectiveness of existing rehabilitation efforts. On the other side of the issue there are the pro-family advocates who argue that positive family contacts should, logically, have a positive affect on the morale of the prisoner and, therefore, on his or her behavior (Davis, 1988). Some of the literature explores the supportive role the family can play by helping the prisoner to cope with the trials and tribulations of day-to-day confinement. Family contact, used as a reward for good behavior is also supported as an alternative, non-coercive means of promoting conformity.

Prisoner homosexuality and sexually-related violence are identified by some authors as major institutional and administrative concerns (Ibrahim, 1974; Roth, 1970). Whatever the motivation, the sensationalistic nature of human sexuality has helped to make them two prominent issues in the area of prisoner-family research. The specific prisoner-family program "remedy" that has been most widely advocated (and despised by others) is conjugal visitation (Goetting, 1982a and 1982b; Hayner, 1972). Numerous studies have been concluded that violence rates and prisoner homosexuality both decrease in prisons where conjugal visitation is allowed (Howser *et al.*, 1984). These studies, like the "family life-criminality" studies, are correlational and must be viewed with extreme caution. Opponents that conjugal visitation programs create administrative problems, security nightmares, and violate socially defined concepts of punishment (Johns, 1971; Balogh, 1966).

Another major and more recent development in the area of prisoners' family research is focused on the imprisoned female (Mawby, 1982; Carlen, 1982; Greening, 1978; Wheeler; 1974). Serious consideration, for example, is now being given to the emotional trauma and hardship experienced by mothers who are separated from their children due to imprisonment (Sobel, 1982; Henriques, 1981; McCarthy, 1980). Much of this research was initiated as part of a greater movement to gain recognition for and insight into the previously ignored female prison population.

The Role of the Prisoners' Families in Rehabilitation, Social Reintegration, and Recidivism

Forming the final link in the continuum of prisoner-oriented literature, "the role of prisoners' families in rehabilitation, reintegration, and recidivism" research investigates the link between the maintenance of strong family ties during imprisonment with successful prisoner rehabilitation, and their subsequent successful family and societal reintegration after their release from prison (Carlson and Cervera, 1991; Schafer, 1991; Adams and Fischer, 1976). There is really very little to differentiate this category from the "family's effect on the prisoner's institutional behavior" category, except that the former tends to be focused on the inevitable release of the prisoner while the latter emphasizes the short-term institutional costs and benefits associated with the maintenance of family ties. The underlying justification for this category is that all prisoners (with a few exceptions) are released from prison at one time or another and that it is probably better to do everything possible to help them rehabilitate than to make them more alienated and frustrated than they were when they were first imprisoned (Kaslow, 1987, p. 352). The maintenance of family ties encompasses a vast array of approaches and perspectives and takes many forms, from regular mail and phone-call privileges to the provision of house trailers for conjugal visitation (Dickinson, 1984).

There are a number of studies that support strong family contact throughout the period of imprisonment in order to ease the period of readjustment to family life after release (Ekland-Olson *et al.*, 1983; Daehlin and Hynes, 1974; Pendleton, 1973). Recidivism prevention, however, forms the underlying philosophical basis for almost every article addressing prisoners' families from the prisoner's perspective (Hairston, 1988; Sinclair and Chapman, 1973). Numerous comparative studies cite the relatively low recidivism rates for prisoner-family program participants. This approach has been criticized for being highly skewed against the programs. Recidivism tests do not account for the individual prisoner's likelihood of recidivism, which can vary considerably. The question asked is: "Is keeping a prisoner, who is very likely to re-offend, free from offenses for 2 years more successful than keeping a prisoner, who is highly unlikely to re-offend, free from offending for 20 years?" In other words, the measure of effectiveness needs to be individually determined rather than collective.

Prisoner-Family Programs

Prisoner-family programs are addressed in almost all of the articles in all of the categories. A separate category has been created here simply to accommodate those articles that make objective assessments of existing prisoner-family programs without any obvious motive. These articles emphasize the benefits to the families, the prisoners, or the institution according to the specific objectives of the program being evaluated. Some articles are institutional reports, while others are more or less published for communication purposes; for the benefit of social workers and other professionals. Many of the reports that I encountered were qualitative in nature, although some of the experimental programs provided fairly extensive quantitative analyses in their summaries.

The primary types of prisoner-family programs discussed are in-prison child-care, marriage counseling, in-prison educational programs for children of incarcerated mothers, conjugal visitation, and combinations thereof (Schiff, 1985; Marsh, 1983; Showalter and Jones, 1980; Wendorf, 1978; Freedman and Rice, 1977; Taylor and Durr, 1977). Most programs receive mixed reactions from prison officials. Economic considerations and the difficulty in determining the efficacy of programs form the fundamental basis of these objections. As discussed earlier, conjugal visitation programs appear to be the most problematic and often face moral opposition as well.

It was interesting to note that most, if not all, prisoners' families program review demonstrated encouraging signs of success. Of course, success is a relative term and is used, in most instances, to refer to the prevention of recidivism. It was also interesting to note that there were also no accounts of program failures.

The Impact of Imprisonment on the Prisoners' Families

Most of the literature in this the largest category of prisoner-family literature area has emerged in the last twenty years (See Table 1). Much of this literature is qualitative in nature and is dedicated primarily to issues like the economic and emotional impact of incarceration on the families of prisoners, with particular emphasis on the children and wives. Stigma, the loss of an important (and often primary) source of income, child-care, and education are all explored in varying amounts of detail (Fishman, L.T., 1988; Davies, 1980; NACRO, 1971).

Some emerging legal arguments are being made about those aspects of incarceration that deny families of prisoners basic human rights and needs (Holt, 1981; Esposito, 1980; McHugh, 1980; Jackson, 1979; Simpson 1979; Haley 1977; Kent, 1975; Palmer, 1972). These articles identify and explores various deprivations related to the loss of family contact. The separation of infants and young children from their incarcerated mothers, loss of custody, family visitation, and the physical inaccessibility of prisons for many families are all addressed (Henriques 1981; Berzins and Cooper, 1982). Some prisoners' family members argue that *they* are the ones who are being punished, despite the fact that they have not committed a crime.

Another interesting argument is that imprisonment perpetuates the very conditions that the "family-deviance" literature suggests perpetuates the conditions that predispose an individual to a life of criminal behavior (Sack, 1977; Moerk, 1973). Numerous studies emphasize the emotional and economic difficulties experienced by the spouses and children of prisoners (Lowenstein, 1986; Curtis and Schulman, 1984; Ferraro *et al.*, 1983). I have been unable to locate a single study detailing the impact of a woman's imprisonment on her husband. However, wives describe how they are forced to submit to humiliating physical inspections by guards before entering prisons, to perform sex acts in public meeting rooms, and how they must make incredible personal sacrifices in order to keep their homes and families in order while their husbands are in prison (Fishman, L.T., 1988, 1987; Lowenstein, 1984)). What becomes clear, in all of these accounts, is that imprisonment truly extends its tentacles beyond the walls of the prison.

Although it could be considered an entirely separate category, prisoners' families demographics are included in this one. Demographics may not be essential to the study of prisoners' families, but they do produce a very graphic, macro-level image of the economic impact that imprisonment has on them as a collective. A detailed demographic study of prisoners' families could produce extremely valuable information (if one were ever to be undertaken) and it could also be combined with existing demographic information to improve our understanding of their economic situation. For example, the economic consequences of single-parenthood are well documented in existing demographic data (See Table 3). Simply knowing what proportion of inmates are married and/or parents will allow researchers to first make reasonably accurate assessments about the number of individuals who are being directly impacted by the imprisonment and then to determine the immediate economic consequences of that imprisonment on both the prisoners' families and on society in general.

Table 3. Selected Population and Income Statistics for Canada, Ontario, and Metro Toronto

Selected Population/ Income (C$)	Canada	Ontario	Metropolitan Toronto
All Families	7,356,170	2,726,735	585,235
Average Income	51,342	57,227	63,736
Incidence of Poverty	13.2%	10.9%	12.4%
Two-Parent Families	6,401,455	2,383,390	489,920
Average Income	54,667	60,846	68,077
Incidence of Poverty	9.3%	7.4%	9.2%
Lone-Parent Families	954,705	342,805	95,315
Average Income	29,060	32,071	35,627
Incidence of Poverty	35.5%	31.6%	29.4%
Male L-P Familes	168,235	59,000	14,500
Average Income	40,792	44,741	48,027
Incidence of Poverty	18.8%	16.1%	16.7%
Female L-P Families	786,470	283,805	80,815
Average Income	26,550	29,437	33,402
Incidence of Poverty	44.7%	40.8%	37.6%

Sources: *Selected Income Statistics*, Statistics Canada, Catalogue 93-331 and *Profile of Census Divisions and Subdivisions in Canada - Part A*, Statistics Canada, Catalogue 95-337.

The incidence of poverty in female lone-parent families is over three times that of two-parent families. Those wives who are wholly or partially dependent on their husbands' incomes are often forced, due to the unforeseeable nature of their separation, to become dependants of the state. It would not be unreasonable, therefore, to conclude, solely on the basis of available demographic data, that the negative economic consequences of imprisoning a husband/father or wife/mother are significant and may be doing more social harm than good.

Prisoners' families demographics, however, are very sparse (Hairston, 1991; Fishman, S.H., 1983, p. 89). Limited information about the gender, age, and marital status of prisoners is available from various government publications. However other essential information about prisoners (and *all* information about their families) is not collected in Canada and is, therefore, unavailable. The impact of this lack of information is profound. Much of the demographic knowledge we, as a society, have about prisoners and their families is conjecture and, therefore, very problematic. I have taken existing demographic information and combined it with smaller samples found in the journal literature in an attempt to manufacture population estimate of prisoners' families in Canada (See Table 4).

Table 4. Estimated Population of Prisoners' Families in Canada, 1990

Admission Type	Female Offender	Male Offender	Juvenile Offender	Spouses	Children/ Parents	Total
Federal Custody	115	4,132	n/a	1,578	4,420	5,998
Provincial Custody	9,209	105,905	4,508	42,224	133,436	175,659
Probation	10,029	48,966	n/a	21,162	74,039	95,200
Total	19,353	159,003	4,508	63,963	211,894	276,857

Sources: Male and female prison populations from: *Juristat Service Bulletin*, Catalogue 85-002, Vol. 10, No. 20, December, 1990. Spousal rates from: *Basic Facts About Corrections in Canada*, Correctional Services of Canada, 1991. Child/parent rates from: McGowan and Blumenthal, *Why Punish the Children?*, 1978, and R. Shaw, *Prisoners Families*, 1992.

Another officially unrecorded, but significant, demographic variable is the longitudinal separation and divorce patterns of couples after entering the "forced-separation-due-to-imprisonment" period. Some very specific, small-sample studies have produced figures that show as many as 75 percent of couples separated and/or divorced soon after their partner was imprisoned. Reports show that imprisoned mothers are disproportionately sole-custody parents without access to temporary guardians for their children (Iglehart and Stein, 1985). There are no studies, that I have been able to locate, that discuss, even in a cursory way, the long term implications of either situation on children in these family situations.

Summary

Prison administrators and bureaucrats reject most prisoner-family programs as ineffective and/or burdensome. The programs appear to be viewed with cynicism and contempt. Prisoners' families programs completely contradict the traditional and popular notions of punishment. Even when they are accepted there is a very strong possibility that they will not enjoy any kind of longevity or institutional support. Prisons are administered by bureaucrats who must have some accountability for their activities to the general public via their government. Prisons have terrible recidivism records. If a prisoners' families program participant is recidivistic, is that a sign that the program has failed or a sign that the entire prison system has failed? The poor longevity rate of prisoners' families programs would suggest that recidivism is attributed to the failure of the program.

For some inexplicable reason the "family life-criminal behavior" literature almost exclusively identifies parents and siblings as family of the offender, while the remainder of the prisoners' families literature excludes parents and siblings in their considerations. Possible explanations include the fact that parents and siblings often lose their "family" status after their children/siblings reach adulthood. Another possible explanation is that the parents and siblings of prisoners are not usually dependent as spouses and children and are not, therefore, as affected by the negative economic consequences of imprisonment. I also suspect that this is also the main reason that research on husbands of prisoners is conspicuously absent. This is a rationale that requires serious re-examination. If the argument is that prisoners' families suffer economically *and* emotionally then these exclusions are hardly justified.

This argument against these exclusions can also be extended to the generally narrow and specifically homophobic definition of "family" that is employed by prison and program administrators, policy-makers, and researchers alike. Using an objective test of "family" in the prison setting is, quite simply, no longer acceptable. Non-traditional families are also being adversely affected by the imprisonment of their loved ones. Homosexual visiting privileges, for example, are still far from a reality. In fact, as noted earlier, many conjugal visitation programs have, as one of their foremost objectives, the reduction and/or the elimination of homosexual behavior. The determination of successful rehabilitation is also strongly linked to the prisoner's successful reintegration into a "stable," nuclear-family environment. Even in regions where "progressive" programs like conjugal visitation are tolerated, sexual relations with anyone other than a legally recognized spouse is still considered simply unacceptable (Schwartz and Zeisel, 1976; Braswell and Cabana, 1975). Unfortunately, the prisoner who openly expresses his or her desire to reconstitute a non-traditional family relationship on the "outside" will be labeled as an unreformed, high-risk parolee and will be treated as such. The myth that non-traditional family relationships with homosexual partners, "brothers," "sisters," or anyone else who is considered "family" by the prisoner (but not by society) cannot be "stable" is pervasive. Society has increasingly accepted definitions of family that are subjective, inclusive, and tolerant. Prisoners' families programs must also begin to reflect this type of inclusiveness.

Finally, prisoners' families need to abandon the social, political, and economic justification models for prisoner-family programs in favor of a more fundamental humanitarian argument. Advocating the institutional functionality of improved family ties for prisoners has the very real potential to backfire. As noted earlier, the program efficacy tests tend to focus on the recidivism rates of former participants. This means that the program is staking its own reputation and future on a highly skewed and flawed measuring instrument (See also Rieger, 1973, pp. 117-118). More importantly, however, it is also supporting the notion that prisoners' families have a significant role (as opposed to "stake") in the rehabilitation and, therefore, recidivism of the prisoner. That argument can be co-opted and used against the prisoner family movement. It is, in fact, another form of the monolithic, positivistic link between the prisoners' families and crime. Somewhat simplified, the argument would go along the following lines: "Parents are responsible for the earlier criminal behavior, and the spouse is responsible for the subsequent criminal behavior." This perspective fails to critically examine the antecedents of crime and criminality. It is, in other words, simply a means of scape-goating prisoners' families instead of focusing on the real problems associated with a disintegrating penal system and a bankrupt retributive philosophy.

Another reason to resist the "institutional functionality of prisoners' families programs" argument is that it moves the locus of consideration away from the prisoners' families. The primary concern is with the prisoner not the prisoners' families. Under the existing institutional functionality paradigm strong family ties are only of value when they are be demonstrably beneficial to the prison. If it were proven, for example, that strong family ties were of infinite emotional value to the children or spouse of the prisoner but of little or no value to the prisoner or to the prison, then the children or spouse would lose out.

The longer I ponder the issues the more I become convinced that those people responsible for the administration of prisons and prison programs have very little idea that prisoners have families or even of how much their decisions affect these peoples lives and, further, that they do not want the situation to change. The government's reluctance to develop a comprehensive demographic profile of prisoners' families is just one case in point. It is, I am sure, justified as an "unfortunate" but inevitable consequence of the differentiation and prioritization process that I alluded at the beginning of this article. I suggest, however, that there are terrific economic incentives for the government to ignore the impact that imprisonment has for those living on the "outside." To admit direct responsibility for someone else's emotional suffering and economic hardship is not an easy thing to do. It is also potentially expensive in a retribution-based society. It may also mean that society will have to revisit its own retribution-based philosophy of punishment and assess whether or not the existing responses within the criminal justice system even remotely satisfy their original objectives.

Notes

*   I would like to gratefully acknowledge the continuing advice, encouragement and support of Professors Leo Davids, Claudio Duran, and Livy Visano.

**   The earliest research, that I have identified, dedicated specifically and exclusively to the topic of prisoners' families is Bloodgood's 1928 study "Welfare of Prisoner's Families in Kentucky."

References

Adams, D. and J. Fischer. 1976. "The Effects of Prison Residents' Community Contacts on Recidivism Rates," *Corrective and Social Psychology and Journal of*

Behavioral Technology, Methods, and Therapy. 22 (4): 21-27.

Arnold, R.A.. 1990. "Processes of Victimization and Criminalization of Black Women," *Social Justice.* 17 (1): 153-166.

Austin, R.L. 1978. "Race, Father-Absence, and Female Delinquency," *Criminology.* 15 (4): 487-504.

Baker, K.J. and K.R.E. McCormick. 1993. *Prisoners' Families: A Bibliography.* Toronto: Unpublished.

Bakker, L.J., B.A. Morris, and L.M. Janus. 1978. "Hidden Victims of Crime," *Social Work.* 23 (2): 143-148.

Balogh, J.K. 1966. "Conjugal Visitations in Prisons: A Sociological Perspective," *Federal Probation.* 28: 52-58.

Barker, G.H. 1940. "Family Factors in the Ecology of Juvenile Delinquency," *Journal of Criminal Law and Criminology.* 30: 681-691.

Bedford, A. 1974. "Women and Parole," *British Journal of Criminology.* 14 (2): 106-117.

Berzins, L. and S. Cooper. 1982. "Political Economy of Correctional Planning for Women — The Case of the Bankrupt Bureaucracy," *Canadian Journal of Criminology.* 24 (4): 399-416.

Bolton, F.G. and J.W. Reich. 1977. "Delinquency Patterns in Maltreated Children and Siblings," *Victimology.* 2 (2): 349-357.

Borgman, R. 1985. "The Influence of Family Visiting Upon Boy's Behavior in a Juvenile Correctional Institution," *Child Welfare.* 64 (6): 629-638.

Braswell, M. and D.A. Cabana, D.A. 1975. "Conjugal Visitation and Furlough Programs for Offenders in Mississippi," *New England Journal on Prison Law.* 2: 67-72.

Briar, K.H. 1983. "Jails: Neglected Asylums," *Social Casework: The Journal of Contemporary Social Work.* pp.387-393.

Buikhuisen, W., C. van der Plas-Korenhoff and E.H.M. Bontekoe. 1982. "Parental Home and Deviance," *International Journal of Offender Therapy and Comparative Criminology.* 201-210.

Burt, M.J.G. 1982. "The 'At Risk' Parolee: Victoria's Special Supervision Unit," *International Journal of Offender Therapy and Comparative Criminology.* 26 (2): 121-125.

Byrne, K. 1976. "Psychodramatic Treatment Techniques with Prisoners in a State of Role Transition," *Journal of Sociology and Social Welfare.* 3 (6): 731-741.

Carlen, P. 1982. "Papa's Discipline: An Analysis of Disciplinary Modes on the Scottish Women's Prison," *Sociological Review.* 30 (1): 97-124.

Carlson, B.E. and N. Cervera. 1991. "Inmates and Their Families: Conjugal Visits, Family Contact, and Family Functioning," *Criminal Justice and Behavior.* 18 (3): 318-331.

Cavan, R.S. and E.A. Zemans. 1958. "Marital Relationships of Prisoners in Twenty-Eight Countries," *Journal of Criminal Law, Criminology and Police Science.* 49: 133-139.

Chaiklin, H. 1972. "Integrating Correctional and Family Services," *American Journal of Orthopsychiatry.* 42 (5): 784-791.

Cobean, S.C. and P.W. Power. 1978. "The Role of the Family in the Rehabilitation of the Offender," *International Journal of Offender Therapy and Comparative Criminology.* 22 (1): 29-38.

Crosthwaite, A. 1975. "Punishment for Whom? The Prisoner or His Wife," *International Journal of Offender Therapy and Comparative Criminology.* 19 (3): 275-284.

Curtis, R.L. and S. Schulman Jr. 1984. "Ex-Offenders, Family Relations, and Economic Supports: The 'Significant Women' Study of the TARP Project," *Crime and Delinquency.* 30 (4): 507-528.

Daehlin, D. and J. Hynes. 1974. "A Mother's Discussion Group in a Women's Prison," *Child Welfare.* 53 (7): 464-470.

Daniel, S.W. and C. Barrett. 1981. "The Needs of Prisoners' Wives: A Challenge for the Mental Health Professions," *Community Mental Health Journal.* 17 (4): 310-322.

Davids, L. "Jews in Prison: The Inmate and His Community," *Journal of Jewish Communal Service.* 62-68.

Davies, R.P. 1980. "Stigmatization of Prisoner's Families," *Prison Service Journal.* 40: 12-14.

Davis, R. 1988. "Education and the Impact of the Family Reunion Program in a Maximum Security Prison," *Journal of Offender Counseling, Services and Rehabilitation.* 12 (2): 153-159.

Eckerd, J. 1988. "Responsibility, Love and Privatization: A Businessman's Guide to Criminal Rehabilitation," *Policy Review.* 45: 52-55.

Ekland-Olson, S., J. Supancic, J. Campbell and K. Lenihan. 1983. "Post-Release Depression and the Importance of Familial Support," *Criminology.* 21 (2): 253-275.

Esposito, S.C. 1980. "Conjugal Visitation in American Prisons Today," *Journal of Family Law.* 19 (2): 313-330.

Ferraro, K.J., J.M. Johson, S.R. Jorgensen and F.G. Bolton. 1983. "Problems of Prisoners' Families," *Journal of Family Issues.* 4 (4): 575-591.

Fishman, L.T. 1986a. "Prisoners' Wives' Interpretations of Male Criminality and Subsequent Arrest," *Deviant Behavior.* 7: 137-158.

Fishman, L.T. 1986b. "Repeating the Cycle of Hard Living and Crime: Wives' Accommodation to Husbands' Parole Performance," *Federal Probation.* 50: 44-54.

Fishman, L.T. 1987. "Patterns of Accommodation Among Wives of Criminals," *Journal of Contemporary Ethnography.* 16: 176-204.

Fishman, L.T. 1988a. "Prisoners and Their Wives: Marital and Domestic Affects of Telephone Contacts and Home Visits," *International Journal of Offender Therapy and Comparative Criminology.* 32 (1): 55-66.

Fishman, L.T. 1988b. "Stigmatization and Prisoners' Wives' Feelings of Shame," *Deviant Behavior.* 9: 169-192.

Fishman, S.H. 1982. "The Impact of Incarceration on Children of Offenders," *Journal of Children in Contemporary Society.* 15 (1): 89-99.

Flanagan, T.J. 1980. "The Pains of Long-Term Imprisonment: A Comparison of British and American Perspectives," *British Journal of Criminology.* 20 (2): 148-156.

Flanagan, T.J. 1981. "Dealing with Long-Term Confinement: Adaptive Strategies and Perspectives Among Long-Term Prisoners," *Criminal Justice and Behavior.* 8 (2): 201-222.

Ford, C.A. 1929. "Homosexual Practices of Institutionalized Females," *Journal of Abnormal and Social Psychology.* 23: 442-448.

Fox, S.S. 1982. "Families in Crisis: Reflections on the Children and Families of the Offender and the Offended," *International Journal of Offender Therapy and Comparative Criminology.* 25 (3): 2254-2648.

Freedman, B.J. and D.G. Rice. 1977. "Marital Therapy in Prison: One-Partner 'Couple Therapy'," *Psychiatry.* 40: 175-183.

Friedman, S. and T.C. Esselstyn. 1965. "The Adjustment of Children of Jail Inmates," *Federal Probation*, 24: 55-59.

Fritsch, T.A. and J.D. Burkhead. 1981. "Behavioral Reactions of Children to Parental Absence Due to Imprisonment," *Family Relations.* 30 (1): 83-88.

Gagnon, J. and W. Simon. 1968. "The Social Meaning of Prison Homosexuality," *Federal Probation.* 32: 23-29.

Gibbs, C. 1971. "The Effects of the Imprisonment of Women Upon Their Children," *British Journal of Criminology.* 11 (2): 113-130.

Goetting, A. 1981. "The Guatemala Prison System: An Application of Familism," *Prison Journal.* 78-81.

Goetting, A. 1982a. "Conjugal Association in Prison: Issues and Perspectives," *Crime and Delinquency.* 28 (1): 52-71.

Goetting, A. 1982b. "Conjugal Association in Prison: The Debate and Its Resolutions," *New England Journal on Prison Law.* 8 (1): 141-154.

Goetting, A. 1984. "Conjugal Association Practices in Prisons of the American Nations," *Alternative Lifestyles.* 6 (3): 155-174.

Goetting, A. and R.M. Howsen. 1986. "Correlates of Prisoner Misconduct," *Journal of Quantitative Criminology.* 2 (1): 49-67.

Graham, F. 1979. "Children In Custody," *Prison Service Journal*, 35: 5-9.

Greening, B. 1978. "A Prison for Mom and Kids," California Youth Authority Quarterly, 31 (4): 25-30.

Griffiths, A.W. and A.T. Rundle. 1976. "A Survey of Male Prisoners: Some Aspects of Family Background," *British Journal of Criminology*. 16 (4): 352-366.

Grogan, H. and R. Grogan. 1968. "The Criminogenic Family: Does Chronic Tension Trigger Delinquency?," *Crime and Delinquency*. 14 (3): 220-225.

Guze, S.B., D.W. Goodwin, and J.B. Crane. 1970. "A Psychiatric Study of Wives of Convicted Felons: An Example of Assortive Mating," *American Journal of Psychiatry*. 126 (12): 1773-1776.

Hairston, C.F. 1988. "Family Ties During Imprisonment: Do They Influence Future Criminal Activity?," *Federal Probation*. 52: 48-52.

Hairston, C.F. 1989. "Men In Prison: Family Characteristics and Parenting Views," *Journal of Offender Counseling, Services and Rehabilitation*. 14 (1): 23-30.

Hairston, C.F. 1991. "Family Ties During Imprisonment: Important to Whom and For What," *Journal of Sociology and Social Welfare*. 18 (1): 87-104.

Hairston, C.F. and P.W. Lockett. 1985. "Parents in Prison: A Child Abuse and Neglect Prevention Strategy," *Child Abuse and Neglect*. 9 (4): 471-477.

Hairston, C.F. and P.W. Lockett. 1987. "Parents in Prison: New Directions for Social Services," *Social Work*, 32 (2): 162-164.

Haley, K. 1977. "Mother's Behind Bars: A Look at the Parental Rights of Incarcerated Women," *New England Journal on Prison Law*. 4 (1): 141-155.

Handler, E. 1974. "Family Surrogates as Correctional Strategy," *Social Service Review*. 48: 539-549.

Hartz-Karp, J. 1983. "The Impact of Infants in Prison on Institutional Life: A Study of the Mother-Infant Programme in Western Australia," *Australian and New Zealand Journal of Criminology*. 16: 172-188.

Hassin, Y. 1979. "The Prisoner's Family in Israel: Problems and Dilemmas," *International Social Work*. 22: 9-16.

Hayner, N.S. 1972. "Attitudes Toward Conjugal Visits for Prisoners," *Federal Probation*. 4 (36): 43-49.

Henriques, Z.W. 1981. "Human Rights of Incarcerated Mothers and Their Children," *International Child Welfare Review*. 49: 18-27.

Hinds, L. 1981. "Impact of Incarceration on Low-Income Families," *Journal of Offender Counselling, Services and Rehabilitation*. 5: 8-9.

Holt, K.E. 1982. "Nine Months to Life — The Law and the Pregnant Inmate," *Journal of Family Law*. 20: 523-543.

Homer, E.L. 1979. "Inmate-Family Ties: Desirable but Difficult," *Federal Probation*. 43 (1): 47-52.

Hopper, C.B. 1962. "The Conjugal Visit at Mississippi State Penitentiary," *Journal of Criminal Law, Criminology and Police Science*. 53: 340-344.

Howser, J.F., J. Grossman, and D. Macdonald. 1983. "Impact of Family Reunion Program on Institutional Discipline," *Journal of Offender Counselling, Services and Rehabilitation*. 8 (1/2): 27-36.

Howser, J.F. and D. Macdonald. August 1982. "Maintaining Family Ties," *Corrections Today*. 44: 96-98.

Ibrahim, A.I. 1974. "Deviant Sexual Behavior in Men's Prisons," *Crime and Delinquency*. 20 (1): 38-44.

Iglehart, A.P. and M.P.Stein. 1985. "The Female Offender: A Forgotten Client?," *Social Casework: The Journal of Contemporary Social Work*. 152-159.

Jackson, J.H. 1979. "The Loss of Parental Rights as a Consequence of Conviction and Imprisonment: Unintended Punishment — A Note," *New England Journal on Prison Law*. 6: 61-112.

Jacobs, J.B. 1976. "A Few Doubts On "Reintegrating" the Offender," *Social Relations*. 11 (2): 191-196.

Johnson, R. 1978. "Youth in Crisis: Dimensions of Self-Destructive Conduct Among Adolescent Prisoners," *Adolescence*. 13 (51): 461-482.

Johns, D.R. 1971. "Alternatives to Conjugal Visiting," *Federal Probation*. 35: 48-52.

Jorgensen, J.D., S.H. Hernandez and R.C. Warren. 1986. "Addressing the Social Needs of Families of Prisoners: A Tool for Inmate Rehabilitation," *Federal Probation*. 50: 47-52.

Karpman, B. 1948. "Sex Life in Prison," *Journal of Criminal Law, Criminology and Police Science*. 38 (5): 475-486.

Kaslow, F.W. 1987. "Couples or Family Therapy for Prisoners and Their Significant Others," *American Journal of Family Therapy*. 15 (4): 352-360.

Kent, N.E. 1975. "The Legal and Sociological Dimensions of Conjugal Visitation in Prisons," *New England Journal on Prison Law*. 2: 47-65.

King, P. 1964. "The Woman Inmate's Contacts with the Outside World," *American Journal of Corrections*. 26: 18-21.

Lanier Jr., C.S. 1991. "Dimensions of Father Child Interaction in a New York State Prison," *Journal of Offender Counselling Services and Rehabilitation*. 16 (3/4): 27-42.

LeClair, D.P. 1978. "Home Furlough Program Effects on Rates of Recidivism," *Criminal Justice and Behavior*. 5 (3): 249-259.

LeFlore, L. and M.A. Holston. 1989. "Perceived Importance of Parenting Behaviors as Reported by Inmate Mothers: An Exploratory Study," *Journal of Offender Counselling, Services and Rehabilitation*. 14 (1): 5-21.

Lochhead, S. 1979. "Prison Visiting," *Prison Service Journal*. 33: 5-6,13.

Lowenstein, A. 1984. "Coping with Stress: The Case of Prisoners' Wives," *Journal of Marriage and the Family.* 46 (3): 699-708.

Lowenstein, A. 1986. "Temporary Single-Parenthood: The Case of Prisoners' Families," *Family Relations.* 35: 79-85.

Marsh, R.L. 1983. "Services for Families: A Model Project to Provide Services for Families of Prisoners," *International Journal of Offender Therapy and Comparative Criminology.* 27: 156-162.

Maskin, M.B. and E. Brookins. 1972. "The Effects of Parental Composition on Recidivism Rates in Delinquent Girls," *Journal of Clinical Psychology.* 30: 239-242.

Mawby, R.I. 1982. "Women In Prison: A British Study," *Crime and Delinquency.* 28 (1): 24-39.

McCarthy, B.R. 1980. "Inmate Mothers: The Problems of Separation and Reintegration," *Journal of Offender Counselling, Services and Rehabilitation.* 4 (3): 199-212.

McCord, J. 1970. "Some Child-Rearing Antecedents of Criminal Behavior in Adult Men," *Journal of Personality and Social Psychology.* 37: 1477-1486.

McHugh, G.A. 1979. "Protection of the Rights of Pregnant Women in Prisons and Detention Facilities," *New England Journal on Prison Law.* 6 (2): 231-263.

McMichael, P. 1974. "After-Care, Family Relationships and Reconviction in a Scottish Approved School," *British Journal of Criminology.* 14 (3): 236-247.

Moerk, E.L. 1973. "Like Father Like Son: Imprisonment of Fathers and Psychological Adjustment of Sons," *Journal of Youth and Adolescence.* 2 (4): 303-312.

Monger, M. and J.A. Pendleton. 1970. "The Nottingham Prisoners Families Project," *Probation.* 84-86.

Monger, M. and J.A. Pendleton. 1977. "The Prison Visit," *Social Work Today.* 8 (34): 8-10.

Morris, P. 1967. "Fathers in Prison," *British Journal of Criminology.* 7: 424-430.

Nacci, P.L., and T.R. Kane. 1984. "Inmate Sexual Aggression: Some Evolving Propositions, Empirical Findings, and Mitigating Counter-Forces," *Journal of Offender Counselling, Services and Rehabilitation.* 9 (1/2): 1-20.

Nash, K.B. 1981. "Family Interventions: Implications for Corrections," *Corrective and Social Psychology and Journal of Behavioral Technology, Methods, and Therapy.* 27 (4): 158-166.

Neussendorfer, J. "Marriage Group-Counseling Inside," *American Journal of Corrections.* 13 (4): 33-34.

N.A.C.R.O. 1971. "Family Visits to Prison: A Survey by N.A.C.R.O.," *Prison Service Journal.* 1: 13-15.

Ouston, J. 1984. "Delinquency, Family Background and Educational Attainment,"

British Journal of Criminology. 24 (1): 2-26.

Page, R.C., M. King, R. Pass and D. Glenn. 1980. "A Comparison of the Parents of Male Inmates with and Without Drug Abuse Problems," *Journal of Offender Counselling, Services and Rehabilitation.* 5 (1): 41-54.

Palmer, R.D. 1972. "The Prisoner-Mother and Her Child," *Capital University Law Review.* 1 (1): 127-144.

Payne, A.T. 1978. "The Law and the Problem Parent: Custody and Parental Rights of Homosexual, Mentally Ill and Incarcerated Parents," *Journal of Family Law.* 16 (4): 797-818.

Pelka-Slugocka, M.D. and L. Slugocki. 1980. "The Impact of Imprisonment on the Family Life of Women Convicts," *International Journal of Offender Therapy and Comparative Criminology.* 24 (3): 249-259.

Pendleton, J.A. 1973. "Through-care with Prisoners and their Families in England," *International Journal of Offender Therapy and Comparative Criminology.* 17 (1): 15-28.

Pimpernell, P. and A. Treacher. 1990. "Using a Videotape to Overcome Clients' Reluctance to Engage in Family Therapy," *Journal of Family Therapy.* 12: 59-71.

Pueschel, J. and R. Moglia. 1977. "The Effects of the Penal Environment on Familial Relationships," *Family Coordinator.* 26 (4): 373-375.

Radelet, M.L., M. Vandiver, and F.M. Berardo. 1983. "Families, Prisons, and Men with Death Sentences," *Journal of Family Issues.* 4 (4): 593-612.

Richards, B. 1978. "The Experience of Long-Term Imprisonment: An Exploratory Investigation," *British Journal of Criminology.* 18 (2): 162-169.

Rieger, W. 1973. "A Proposal for a Trial of Family Therapy and Conjugal Visits in Prison," *American Journal of Orthopsychiatry.* 43 (1): 117-122.

Roth, L.H. 1971. "Territoriality and Homosexuality in a Male Prison Population," *American Journal of Orthopsychiatry.* 41 (3): 510-513.

Sack, W.H. 1977. "Children of Imprisoned Fathers," *Psychiatry.* 40 (2): 163-174.

Sack, W.H. and J. Seidler. 1978. "Should Children Visit Their Parents in Prison?," *Law and Human Behavior.* 2 (3): 261-266.

Sack, W.H., J. Seidler and S. Thomas. 1976. "The Children of Imprisoned Parents: A Psychosocial," *American Journal of Orthopsychiatry.* 46 (4): 618-628.

Sametz, L. 1980. "Children of Incarcerated Women," *Social Work.* 25 (4): 298-302.

Schafer, N.E. 1978. "Prison Visiting: A Background for Change," *Federal Probation.* 42 (2): 47-50.

Schafer, N.E. 1991. "Prison Visiting Policies and Practices," *International Journal of Offender Therapy and Comparative Criminology.* 35 (3): 263-275.

Schiff, S.S. 1985. "Children of Incarcerated Parents: The Preschool in Prison Project,"

Early Childhood Education. 18 (1).

Schneller, D.P. 1976. "Conjugal Visitation: Prisoners' Privilege or Spouse's Right," *New England Journal on Prison Law*. 2 (2): 165-171.

Schwartz, M.C. and J.F. Weintraub. 1974. "The Prisoner's Wife: A Study in Crisis," *Federal Probation*. 38: 20-26.

Schwartz, M.C. and L. Zeisel. 1976. "Unmarried Cohabitation: A National Study of Parole Policy," *Crime and Delinquency*. 22: 137-148.

Shaw, S., 1990. "A Service for Prisoners' Families," *Prison Service Journal*. 80: 33-35.

Shoham, S.G., G. Rahav, *et al*. 1987. "Family Parameters of Violent Prisoners," *Journal of Social Psychology*. 127 (1): 83-91.

Showalter, D. and C.W. Jones. 1980. "Marital and Family Counselling in Prisons," *Social Work*. 25: 224-228.

Simpson, C.1979. "Conjugal Visiting in United States Prisons," *Columbia Human Rights Law Review*. 10: 643-669.

Sinclair, I. and B. Chapman. 1973. "A Typological and Dimensional Study of a Sample of Prisoners," *British Journal of Criminology*. 13 (4): 341-353.

Singh, R.G. 1978. "Prisoner Adaptation of Surrendered Dacoits," *Indian Journal of Social Work*. 39 (1): 27-31.

Sobel, S.B.1982. "Difficulties Experienced by Women in Prison," *Psychology of Women Quarterly*. 7 (2): 107-118.

Sterling, J.W. and R.W. Harty. 1972. "An Alternative Model of Community Services for Ex-Offenders and Their Families," *Federal Probation*. 36: 31-34.

Stollery, P. 1970. "Families Come to the Institution: A 5-Day Experience in Rehabilitation," *Federal Probation*. 34 (4): 46-53.

Taylor, H.L. and B.M. Durr. 1977. "Preschool in Prison," *Young Children*. 32: 27-32.

Thomas, R.G. 1981. "The Family Practitioner and the Criminal Justice System: Challenges for the 80's," *Family Relations*. 30: 614-624

Walker, N. 1983. "Side Effects of Incarceration," *British Journal of Criminology*. 23 (1): 61-71.

Walker, W. 1972. "Games Families of Delinquents Play," *Federal Probation*. 36: 20-24.

Weathers, L. and R.P. Liberman. 1975. "Contingency Contracting with Families of Delinquent Adolescents," *Behavior Therapy*. 6: 356-366.

Weintraub, J. 1976. "The Delivery of Services to Families of Prisoners," *Federal Probation*. 40: 28-31.

Wendorf, D.J. 1978. "Family Therapy: An Innovative Approach in the Rehabilitation of Adult Probationers," *Federal Probation*. 42 (1): 40-44.

Wheeler, M. 1974. "The Current Status of Women in Prison," *Criminal Justice and Behavior.* 1 (4): 371-380.

Wilcoxin, S.A. 1986. "One-Spouse Marital Therapy: Is Informed Consent Necessary?," *American Journal of Family Therapy.* 14: 265-270.

Wilson, G. 1984. "I Know While He is in Prison He's Safe," *New Society.* 70 (1): 172-174.

Wilson, H. 1975. "Juvenile Delinquency, Parental Criminality and Social Handicap," *British Journal of Criminology.* 15 (3): 241-250.

Wilson, H. 1980. "Parental Supervision: A Neglected Aspect of Delinquency," *British Journal of Criminology.* 20 (3): 203-235.

Wilson, H. 1982. "Parental Responsibility and Delinquency: Reflections on a White Paper Proposal," *Howard Journal.* 21: 23-34.

Worthy, A. 1975. "The Probation Service and Matrimonial Reconciliation," *International Journal of Offender Therapy and Comparative Criminology.* 19 (3): 270-274.

Zeitz, D. 1963. "Child Welfare Services in a Women's Correctional Institution," *Child Welfare.* 42: 185-190.

Zemans, E. and R. Cavan. 1958. "Marital Relationships of Prisoners," *Journal of Criminal Law, Criminology and Police Science.* 49 (1): 50-57.

Community-Based Policing:
The Rhetoric of Community Participation

A Case Study

Shivu Ishwaran

> ...[E]xpansion, dispersal, invisibility, penetration — is indeed continuous — with those original transformations. The prison remains — a stubborn continuous presence, seemingly impervious to all attacks — and in its shadow lies 'community control.'
> Stanley Cohen, *Visions of Social Control* (1985: 85).

Introduction

Recently the concept of community participation has become transformed as an attractive contrivance of organizational discourse. Throughout the last few decades Canadian institutions have advanced a proliferation of programs, strategies and policies apparently designed to elicit a greater degree of "community" participation. This new ideology depicts the continuation and intensification of the "carceral" apparatus, manifested as viable alternatives of social control located in the community. To elaborate, community participation, expressed through the rhetoric of community-based policing, has been sought to supplement traditional and more formal methods of organizational control. This new ideological movement has attained a heightened significance within the "chatter" of control (Foucault, 1977: 304).

The security industry, as an expression of the behavior of control, has been busy investing resources in the "community" to enhance a greater degree of co-operation, input and control. As early as 1970, the burgeoning growth in private police was roughly equivalent to that of their public counterparts (Shearing and Stenning, 1983). In Ontario, estimates of contract security from 1967 through 1975 indicate a growth rate greater than that of the general population and double the growth rate of the public police. This so-called "quiet revolution of policing" has infiltrated deep into the social fabric of Western society, into and beyond the exclusive domain of the public police (Shearing, Farnell and Stenning, 1980).

In fact, private security is increasingly engaged in the maintenance of public order because more and more public life is conducted on privately owned and controlled property, what Stenning and Shearing call "mass private property" (shopping centers, residential estates, campuses, airports) — areas from which the public police are restricted from routine access due to legal constraints and inadequate resources.

Researchers and practitioners of private security have had little problem defining the phenomenon. Shearing and Stenning (1983: 3) offer a general definition of private security:

> Private security refers to the process whereby individuals and agencies (be they governments or corporations) make use of the age-old perogative of self-help to protect information, property and persons.

However, defining the boundaries of private security has not been so easy. Security forces may be categorized in several ways: by who employs them — a public agency or a private business, institution, or individual; by the degree of police powers they possess; or by the functions they perform. There are a number of quasi-public and quasi-private forces which occupy the gray area between these two extremes. Confusing the matter further are the blurred distinctions between private property and public property.

The purpose of this paper is to evaluate critically the concept of community by providing a case study of a much celebrated exercise in social control — community policing — and to focus attention on what Visano (1993) identifies as the need to transform community inaction into communities-in-action. The case study evaluates a large Canadian university's private and community-based security from the students' perspective.

Much is known about the efforts of private security in infiltrating, penetrating and forming connections to local initiatives. Yet relatively little information exists documenting the extent to which the community actually participates in its own right, *sui generis*, in organizational activities.

That is, a paradox is apparent regarding outside community initiatives. What happens when, for example, representatives of community-based organizations, like the university student bodies, demand an agenda that departs from administratively-filtered priorities? For instance, a discussion of sexual harassment indicates that many members of the university community have requested, on the part of the complainants, to make some situations of sexual assault and harassment public and political. This often is in conflict with the university's polices and procedures which state clearly that confidentiality is necessary for the protection of the complainant and the accused. Although confidentiality is an important component of most sexual harassment complaints, there are circumstances under which the population feels that the procedures which have been set up to protect them serves only to "keep things quiet" and goes against their wish to have the issues made public. There is nothing in current procedure or policy which addresses this new and increasing pressure to politicize sexual assault and harassment issues. Consequently, campus security is facing similar pressures as the public police, to become more accountable to the community which they serve and to re-focus their policing approach to meet the community's needs. Recently, one university's Security Advisory Committee (1991) reported a change in philosophy, including a move to a community-based, private policing and security service for the university community. This form of private policing is viewed as part of a wider informal justice and community control movement, a wish to return to "basic" familiar values (Scull, 1977: 41).

The term "community-based policing" is historically founded in the idea that when a community looks into a mirror, they should see their police force, and conversely, that when a police force looks into a mirror they should see their community. It is very necessary that there be a closer kinship between the community and the police, that they both be on the same level and that they try to reflect the image of the people themselves.

The basic philosophy of community policing proposes a preventive approach to crime. This philosophy holds that no police work can be carried out without a high level of communication and understanding between the police and the public. Crimes cannot be prevented or solved without information from the public. If the true test of police efficiency is found in the absence of crime and disorder and not in the visible evidence of the police in dealing with it, then community policing must focus on prevention, referral and educational functions and planned strategies designed to have an impact on problems prior to their occurrence. Changing public service expectations to community consultation and education potentially allows the police to more effectively and efficiently deliver police services.

Because communities lack complete information about policing problems, information provided by the police can help establish more informed and rational expectations. Unless the police themselves take the initiative in sharing this information, community problem identification will be at best incomplete and more often simply inaccurate or ill-informed. A proactive management approach means harnessing information analysis and the problem-solving abilities of the police as part of a more active and educational role for community policing.

Objective of the Study

As stated earlier, the purpose of this study is to evaluate critically the concept of community by providing a case study of private and community policing and to focus on the extent to which the community actually participates in its own right, *sui generis*, in organizational activities.

Student participation is operationalized in two ways, first, by measuring students' awareness of, and participation in, crime prevention information and practices. These security programs and related information services furnish avenues of community participation as they represent channels of communication between the University's administration, campus security and the student body, as well as providing students with the opportunity to participate in their own safety.

Second, although students can simply indicate "yes" or "no" to awareness questions, this measure does not provide an indication of whether they feel adequately informed. Thus, students are asked to rate the effectiveness of campus security and to rate their university's efforts to inform them about the availability of various security measures.

Methodology

The sample for this study comprises a non-random selection of 246 undergraduate students at a Canadian university. Of the respondents, 61.7 percent (n=152) are students enrolled in a third-year sociology course relating to issues of crime and delinquency. The decision to study this particular class was based upon the relevance of the course content to this study. For example, topics covered include: community corrections; definitions of community; community participation; rhetoric of "community" control and discipline; power and authority conflicts; bureaucracy and policing and the organizational nature of policing and victimization. It is argued here that because students in this class are sensitized to a large variety of macro and micro perspectives in criminology, they are likely to be aware of issues related to crime and crime control relevant to their own lives and environment. If students do not demonstrate such an awareness to issues of crime on campus, this represents a significant result. The remainder of students surveyed, 38.2 percent (n=94), are enrolled in a first-year introductory sociology course. This course was chosen for study in order to provide for the representation of first- and second-year students. Of the students surveyed, 71.49 percent are of an average of 18 to 22 years of age.

Of the respondents, 80.49 percent (n=198) are female and 19.51 percent (n=48) are male. The under-representation of males is not considered detrimental to the study because most of the security issues, practices and programs are aimed at female students.

The distribution by year of study indicates that 30.08 percent of students surveyed are in their first year, 25.20 percent in their second year, 32.11 percent in their third year and 10.98 percent in their final year of study.

The primary mode of data collection is based upon a group-administered survey, titled "Questionnaire about students awareness of campus security issues." Respondents were asked to indicate "yes," "no," or "don't know" to questions regarding their awareness of various security measures and crime prevention education services offered by their university. These services included: escort, emergency phones, self-defense courses, preventive workshops and lectures, support organizations such as the sexual harassment education and complaint center, peer support center, counselling and development center, womens' center and campus crime brochures. Students were asked to rate on a five-point Likert response format (very poor/poor, satisfactory, good/very good) their opinion of the effectiveness of campus security and to rate their university's efforts to inform students of the availability of those security services.

Results[1]

Escort Service

Table 1 summarizes the overall ratings students gave the campus escort service.[2] As shown here many respondents rated negatively their university's efforts to inform them of the service. This finding is also intriguing, and somewhat paradoxical in light of the fact that 94.3 percent of students indicated at least an awareness of this highly visible service. This paradox again emerges in terms of two other related questions.

Table 1. Rating of University's Efforts to Inform Students of the Following:

	poor/ very poor	satisfactory	good/ very good
Escort Service	38.2%	31.3%	26.4%
Emergency Phones	54.9%	24.8%	16.7%
Self-defence Course	72%	18.7%	2.9%
Support Organizations	30.5%	35.4%	28.9%
Lectures/Workshops (prevention of crime)	50.8%	33.7%	9.8%
Crime on Campus	48.3%	34.2%	13.4%

The first has to do with how students responded to the question "How effective is the escort service?" Summarized in Table 2 are the students' overall sentiments with regards to this question. Indeed, the responses presented here would indicate that the majority of the students "don't know," being either unsure or ill-informed about the effectiveness of the service. In the second, students were asked, in an open-ended question, to comment on their response to the previous question. In 40.3 percent of the comments provided, students revealed that they had never, in fact, used the service, another 31.9 percent responded with a general negative comment and 15.3 percent indicated a lack of information regarding campus escort procedures. Only 0.5 percent provided any sort of positive comments at all. In summary, then, we see that 87.5 percent of the comments indicated that students had either never used the service, had in some way commented negatively on it or had lacked adequate information to make any kind of an informed judgement on it.

Table 2. Students' Rating of the Effectiveness of the Escort Service

	poor/ very poor	satisfactory	good/ very good	don't Know
Escort Service	13.8%	15%	8.5%	58.5%

Thus, although a significant majority of students indicated at least an awareness of the escort service, they lacked any sort of substantive understanding conducive to its being more effectively utilized. Students' adverse rating of their university's efforts to inform them of the escort service is apparently a consequence of their lack of adequate information regarding its role and function on campus.

Emergency Telephones

As can be seen in Table 1, the majority of students expressed a negative opinion of their university's efforts to properly inform them of the emergency telephone service available on campus. Table 3 summarizes data on three additional questions asked in relation to this service. Results obtained on the first question indicate that the majority of students surveyed were at least aware of the phones. The second question revealed, however, that while this was the case, half of these students did not know how to identify the phones. On the third question, a majority of the students expressed, moreover, that they were not comforted by the presence of the phones.

To sum up, then, we see a paradox again emerging in the students' responses; for although they did indeed, indicate their awareness of the service, it is argued here that students' ambivalent and negative opinions of their university's efforts to adequately inform them was a result of their lacking appropriate information which would enable them to better identify the service.

Table 3. Questions About Emergency Telephones

	Yes	No/ Don't know
Are there emergency phones on campus?	**75.2%**	**24.8%**
Do you know how the phones are identified?	**50%**	**50%**
Does the presence of emergency phones make you feel safer?	**35.8%**	**62.6%**

Self-Defense Course

The data presented in Table 1 show that the majority of students responded negatively to the question of whether or not they believed their university had adequately and properly informed them of the Wen-do self-defense course offered on campus. Moreover, results obtained in Table 4 indicate that a majority of students are not even aware of the course. It would again seem to support the contention postulated earlier, that students' adverse rating of campus efforts to inform them of this course is itself a reflection of students' lack of appropriate procedural information.

Support Organizations

With regards to campus support organizations we find similar patterns of opinion emerging. As Table 1 shows, students rating of their university's efforts to provide adequate information was evenly distributed among these categories, these being "poor/very poor," "satisfactory" and "good/very good." Table 5 contains data on student awareness of various support organizations and reveals that a significant majority of students are indeed aware of the organizations. Table 4, however, presents the various programs provided by the support organizations and shows that a large percentage of students are not aware of the programs offered. Moreover, when students were asked if they had attended

Table 4. Students' Awareness of the Following Programs:

	Yes	No/ Don't know
Self-defence Course	31.3%	68.3%
Street Proofing Workshops for Women	9.4%	69.8%
Informal Lectures/Workshops Dealing with Violence, Sexual Assault/Date Rape	41.1%	58.5%
Alcohol/Drug Awareness Programs	29.7%	69.5%

any of the programs, 94 percent indicated they had not. Students were largely divided, then, in their attitudes towards university efforts to inform them of the support organizations, and while a majority are aware of these organizations, they are largely unaware of the programs provided by those organizations 94 percent of students not having participated in any of the preventative and informational programs. This again suggests a correlation between students' possession of inadequate information and their participation.

Lectures/Workshops

The university provides a series of lectures and workshops dealing with issues related to the prevention of violence, sexual assault and acquaintance rape, and conflict resolution. Again, we see students' dissatisfaction with university efforts to adequately provide its students with information relating to these programs. In Table 4 we see that a majority of students are unaware of the availability of these lectures and workshops on campus.

Table 5. Students' Awareness of the Following Support Organizations:

	Yes	No/ Don't know
Sexual Harassment Education and Complaint Centre	**80.5%**	**19.5%**
Student Peer Support Centre	**80.1%**	**19.9%**
Counselling and Development Centre	**86.2%**	**13.8%**
Student Affairs	**93.5%**	**6.5%**
Women's Centre	**73.2%**	**26.8%**

We find similar results again with student ratings of the university's reporting of crime on campus (Table 1). In addition, when students were asked if they were aware of the availability of brochures dealing with crime, crime prevention and support services on campus, 60.6 percent indicated they were not.

Lastly, we can observe similar negative opinions in student rating of the effectiveness of campus security the majority (55.7 percent) indicating that they were "unsure," 27 percent responding "ineffective/very ineffective," while only

14 percent expressing that they believed security on campus was "effective/very effective."

Conclusion

The concept of community has become a negotiable commodity, the value of which is conveniently determined by the university's administration. To date, the university's recent change in philosophy, to a private/community-based security structure, represents merely a change in rhetoric, apparently designed to provide ideological legitimacy. As mentioned earlier, the basic tenets of community-based policing are community participation, a preventive approach to crime, and high levels of communication and understanding between the police and the public. The results of this study indicate that student participation, let alone an understanding and awareness of preventive safety matters, is woefully lacking. Students are not satisfied with the administration's efforts to inform them about many of the available security practices, and the community demonstrates a lack of comprehensive knowledge regarding their campus security. In addition, the university's security continues to function within the reactive approach involved in the traditional crime-fighting role, responding to calls for service, imposing the after-the-fact perspective of police work.

However, the community concept provides more than ideological legitimacy. Rather, the concept appears to be designed to discipline outside participation, pre-empt criticism and discourage much-needed critical dialogue. Within this facade of community participation, the student community has limited options since the outside involvement is a creation of the organization.

Community participation as a viable and complex script involves more than reading and rehearsing well-prepared situational roles. Participation occurs within wider interactive contexts and articulates discourses of power and privilege. The appropriation of community resources by the university through its security practices legitimates programs, re-socializes university communities and discredits discordance. Meaningful dialogue, based on a comprehensive knowledge base, is lacking. The security department often seeks a banal accommodation to bureaucratic propaganda rather than the capacity of the community to "advise." Remedial palliatives like the reinstating of ill-informed student participation are shallow gestures that fail to confront structural impediments and erode the maintenance of dependency relations. Students' safety continues to be dependent on the discretionary whims of the administration which holds the prospect of restoring community confidence.

Organizational analyses clearly suggest that bureaucracies are designed to maintain stability while concurrently generating limited outside input. Centralization protects the distribution of knowledge. Controls in decision-making and policy formulations are deliberately complex and blurred, thereby defying facile access to and understanding of the vagaries of administrative privileges. In unmasking authority structures it is evident that the work in private/community-based policing is clouded in secrecy. Decisions, policies and strategies are effectively insulated and immune from general inspection. Secrecy is rationalized, in turn, as an organizational imperative. The norm of secrecy is a valuable tool in controlling information and avoiding accountability. Crime knowledge or the production of occurrence reports is a screen behind which information is protected. Keeping secret their expertise and motives, organizations treat knowledge as a very powerful commodity which is differentially distributed not only within the bureaucracy, but in the wider communities. Experts are assigned exclusive tasks. Mysteries are perpetuated. Specialization dislocates and subordinates public input. Once the public has succeeded in participating in formal discussions, a further institutional layer surfaces — informal occupational cultures that do not necessarily share the political enthusiasm of community involvement. With community security/private policing, the rank-and-file seeks to protect its own control, self-interest and immunities from the incumbrances of management.

The occupational culture arguably has reason to suspect management-driven initiatives such as community policing. Labor is seldom consulted in the wholesale array of impression management schemes that promote the progress and success of administrative plans. Security operations are designed to be appendages of the administration. Alternatives that depart from co-optation are needed despite the resistance from bureaucrats who continue to act with impunity in disregarding the interest of constituencies in favor of their own organizational and political exigencies. As Cohen (1985: 44) suggests, "community control has supplemented rather than replaced traditional methods."

An examination of the formation of equitable bases of knowledge, advocacy and community-based empowerment provides a conceptually more comprehensive appreciation of community action in crime control. From a public policy perspective, however, a focus on "communities-in-action" is threatening. This commitment to meaningful action does not suffer from the vagueness and vulnerability of institutionally-sponsored "community" constructions. Changes in administrative legislation, rules and regulations that protect independent community input are long overdue. Moreover, vigilance on the part of the community groups like the student body in reclaiming that which more appropriately belongs to them is also called for.

The community is an elusive concept that has been too easily appropriated by the state, administrations and organizations to engineer support for limited initiatives that fail to grapple with fundamental inequalities in the access to knowledge regarding private/community-based policing. "Private" policing is contextually determined and discursively constructed to satisfy organizational interests. Without reference to the context of power, the community concept has become a pretext for intervention and exclusion. A commitment to local contexts, for example, is perceived as counter-hegemonic and subject to coercive measures. This sponge-like term enables the organization to celebrate and parade representations that it has effectively screened — to appoint those individuals and organizations who subscribe deferentially to authority/subject relations and enjoy the benefits of such complicity and deception under the guise that something is being done "for" and "with" the community.

Admittedly, this case study of the involvement of student communities suffers from oversimplification and remains suspiciously idiosyncratic and exploratory. Nevertheless, there are generic principles that are readily applicable to other research sites and that demand a more rigorous investigation. This brief discussion urges students of control to juxtapose the rhetoric inviting community input with the actual content and structure of community involvement — the imposed limitations that silence the voices of those concerned. Given the proliferation of chatter about increased community participation, students are seldom encouraged to articulate and problematize the relationship between security practices, safety and democratic accountability.

Lamentably, student community inaction is rewarded by the exaggerated privilege of being permitted to sit on committees struck by administrative functionaries. Communities-in-action, however, mobilize, advocate and articulate an agenda that provides an ongoing critique of power. Inequalities in policing knowledge are legion. Students, especially victims of assault, feel ignored by an alienating system of private justice. Students, as well as their friends and families, suffer deprivation; they complain about the insensitivities of management, stress and poor learning environment conditions; and the general student body remains ignorant and fearful of "alarming" crime rate statistics that are often advanced to secure support for policing practices. Irrespective of the substantive site, university actors, from faculty to students, are advised to redefine their respective role expectations of "their" security and operationalize more fully the meanings of community.

Endnotes

1. Data compilation and analysis was achieved by the use of MYSTAT, an interactive statistical and graphics package. This software can be used on IBM PC-compatible, Vax/VMS, and Macintosh computers. MYSTAT's capacity is 50 variables and 32,000 cases.

2. For organization purposes, the data for each security practice in Table 1 will be dealt with individually to allow for the presentation of other related findings.

References

Boostrom, R. and J. Henderson. 1983. "Community Action and Crime Prevention: Some Unresolved Issues," *Crime and Social Justice*. 19: 24-30.

Cohen, S. 1985. *Visions of Social Control*. Cambridge: Polity Commissioner's Directive, "Citizens' Advisory Committee," January 1, 1987.

"Community Policing in the 1980's: Recent Advances in Police Programs." 1987. Ottawa: Solicitor General of Canada.

Doob, A. and J. Roberts. 1988. "Public Punitiveness and Public Knowledge of the Facts: Some Canadian Surveys," in Walker, N. and M.Hough (eds.), *Public Attitudes to Sentencing: Surveys From Five Counties*. Aldershot: Gower.

Foucault, M. 1977. *Discipline and Punish: The Birth of the Prison*. N.Y.: Pantheon.

Jeffries, Fern. 1977. *Private Policing: An examination of In-House Security Operations*. Toronto: Center of Criminology, University of Toronto.

Kakalik, J.S. and S. Wilkhorn. 1977. *The Private Security: Security and Danger*. New York: Crane, Russak and Co.

Murphy, C. and G. Muir. 1985. *Community-Based Policing: A review of the Critical Issues*. Ottawa: Communications Group.

Shearing, C.D. and P.C. Stenning. 1987. *Private Policing*. California: Sage Publications.

Shearing, C.D. and P.C. Stenning. 1983. *Private Security and Private Justice*. Montreal: Institute for Research on Public Policy.

Spitzer, S. and A.T. Scull. 1977. "Privatization and Capitalist Development: The Case of The Private Police," *Social Problems*. 25: 18-29.

Stenning, P.C. 1975. *Legal Regulation and Control of Private Policing and Security in Canada*. Toronto: University of Toronto.

Visano, L. June 1993. "Community, Control and Contradictions." Symposium: Diverse Sociologies. York University.

Webber, D. 1987. "Community-Based Corrections and Community Consultation — A How to Manual." Solicitor General: Ontario Region.

The Culture of Capital as Carceral: Conditions and Contradictions

Livy A. Visano

Introduction: Beyond the Text

The study of control is an ideological exercise in discredit, an inquiry into confrontations with cultural conformities. The enterprise at hand invites a more intrepid, circumspect and perhaps a more ambitious interrogation of our familiar, traditional as well as common-sense interpretations in forging more critical analyses of penology. Social control is a serious substantive site for investigating often overlooked fundamental issues of inequality, for unravelling the connectedness of concepts and applied practices, and for questioning dominant modes of discourse. A number of issues remain unresolved despite the proliferation of different canonical texts. Notably, in the sociology of crime the quintessential debate concerning the relationship between structure and agency continues to loom large in all conventional encyclopedic overviews. To what extent does structure shape and in turn is shaped by the nature of human agency? Contrary to traditional texts, we are not supporting the fallacies of either determinism or subjectivism — an emphasis on the duality of objects and subjects. This binary method of analysis obfuscates and erroneously reduces subjects (actors) to objects (structures) or alternatively, objects to subjects. Instead of bifurcating the phenomenon of crime in terms of an either/or framework, a priority of agency or structure, it is conceptually more fruitful to examine, as Marx (1956) suggested, the "totality," the interconnectedness of events and activities, the intersections of history, culture and political economy. Throughout Marxist writings, there is no attempt to deny the capability of the human agency in intervening in events. Clearly, structure emerges out of social interactions. Since interactions are "social," they are essentially interconnected and conditioned. Just as the processes of interrelationships among actors constitute structures, structures simultaneously constitute agency. In other words, structures and processes are shaped conjointly, as indicated by Marx

(1956: 74-75) in the following excerpts:

> The production of ideas, conceptions and consciousness is at first directly interwoven with the material activity and the material intercourse and thought, the mental intercourse...still appear at this stage as the direct emanation of their material behavior. The same applies to mental production as it is expressed in the political, legal, moral, religious and metaphysical language of a people.

> People, for Marx, are the producers of their conceptions or ideas — real active agents, as they are conditioned by a determinate development of their productive forces, and of the intercourse which corresponds to these, up to its most extensive forms. Marx's analysis focused on real, active and creative beings, who "from their real life process show the development of the ideological reflexes and echoes of this life-process." (*ibid.*).

Culture, Control and Crime

Culture is a central aspect of society. Social life is conditioned culturally. Crime and punishment occur within cultural contexts, temporally and spatially. Images of control incorporate the manipulation and commodification of aspects of a given culture, general and specific.

More precisely, interactions originate between subjects through meaningful relationships with objects of their environments. In other words, actors and activities are "situated" within certain identifiable settings which contextualize relations and encounters. Unfortunately, traditional texts typically de-contextualize both specific and generic conditions of crime. According to these myopic accounts, crime is advanced as habitualized or institutionalized, assuming a life independent of its constituent elements. A more prudent approach, however, suggests that crime is a consequence of relations of control. Crime, as a localized script, is written within a larger narrative on the appropriateness of control. In fact crime is a challenge that signifies subversion — a defiance of normative and integrative efforts ostensibly designed to "conform" actors, groups or communities. Alternatively, control shapes crime. Crime acts symbolically as a rationale for control and becomes conveniently incorporated within a discourse that supports the intrusions of the more powerful. Power ultimately exploits images of crime. For example, local crusades against prostitution, lyrics in rap music, as well as more global campaigns such as Operation Desert Storm in the Persian Gulf crisis, the struggles in Bosnia, Operation Restore Hope in Somalia, the belligerence of the South African state, the failed coup in the Soviet Union, the provocative gestures of the American intelligence community and the corresponding adventurism of the US military, to name only a few instances, involve much more than the maintenance of tolerable boundaries of diversity. The declarations of deviance by moral entrepreneurs who have routine media access succeed in manufacturing excitement, hysteria and crime waves which are then used to justify further monopolistic interventions. Social control is not only calibrated to contain crime but more importantly, to create public outrage within a morality play that enhances the legitimacy of power. Rhetorical devices appeal to national security and a sacred morality. Interestingly, crime is manipulated. Witness, for instance, the dramatically contradictory types of responses on the part of American government *vis-à-vis* the overthrow of the Allende regime in Chile in the early 1970s and the support generously provided to Boris Yeltsin's regime in Russia in the early 1990s; the American support for the Contras in Nicaragua, or the drive-by bombings on Libya authorized by American President Ronald Reagan in the 1980s. Slogans about democracy and freedom conceal cultural contradictions. Likewise, police forces are inclined to release alarming crime statistics whenever police operational budgets are subjected to careful civilian scrutiny.

In problematizing the relationship between control and culture, the following questions need to be addressed: what are the mechanisms through which the dominant class maintains its hegemony? How do actors consent, resist or accommodate, that is, 'make sense' of power relations? Informed by Marxist writings, we analyze the construction of crime by examining the layered carceral contexts in reference to culture, political economy and the state on the one hand, and subjectivity and consciousness on the other.

Clearly, culture pervades society. Culture, as will be argued, is an ideological superstructure that is reflective and derivative of socio-economic structures. Also, as a social accomplishment culture is rooted in history and expressed in action, forms of consciousness and forces of change. Culture, simply put, is the channel through which social relations are conducted. As a set of shared meanings, expectations and understanding, culture is manifested in symbolic communication — language, customs, myths, signs as well as material artifacts. Culture consists of ideas that are selectively communicated, believed and legitimated quite often as knowledge. Oversimplified threats of danger as well as exaggerated claims of merit are manipulated in order to dismiss, counter or even create different beliefs.

Cultural reproductions are integrally related to political economy and the state. Culture, organized on hegemonic principles, is crucial to the political economy, which in turn depends on instruments of authority to maintain control over capital and profit.

Materialism, Media and Manufactured Margins

Material consumption operates in the realm of ideas, thereby depoliticizing explosive class relations. Ideas are appropriated and distributed by dominant class interests. Within the cultural sphere, there is considerable deference to social engineers who, as "experts," translate and advance the material needs especially of multinational corporations. Credentials, technology and corporatism generate informational capital and cultivate a general consumption that relies on the compliance of the marginalized. Essentially, everyday life is colonized by corporate capitalism. Corporate capital propagates consumption, narcissism and material fetishes not as rewards in affluent societies but essentially as continued investments in a consuming and producing force. The consumer culture is directed by those who profit; materialism exploits social relations.

As noted above, culture is a social production that is inextricably tied to political and economic concatenations of power. It is precisely because the content of the culture of capitalism is material that materialism is assigned primacy. Materialism embodies both the morphology and content of culture.

The media of communication are overwhelmingly clear in scripting a language configured in seductive symbols urging an obsessive deference to materialism. Thus, materialism succeeds in becoming personalized and possessed. Loyalties to this pervasive value system are passionately defended and generationally anchored. All mainstream institutions are implicated in reproducing this morality of individual possessiveness. These normative notions about the status quo are protected and become social catechisms that define the appropriateness of identity and behavior.

Central elements of the "culture industry" — the media, education, arts, entertainment etc. — legitimate the interests of advanced capitalism. As Adorno and Horkheimer note, the "stronger the positions of the culture industry become, the more summarily it can deal with consumers' needs, producing them, controlling them, disciplining them..." (1989: 181). The culture in which we live is, as Itwaru (1989a: 2) clearly argues,

> an escalation and fusion of technology and capitalism — technocapitalism — whose hegemonizing momentum produces deception in a delirium of vertiginous frenzy in the deepening of its control over its human subjects who have in many ways been

persuaded they are in the age of the nirvana of progress.

The media reward consumers and punitively impoverish those actors and/or institutions actively engaged in interrogating the fundamental ethos of passivity, individuality and consumption. The subliminal seduction manipulated (Key, 1981: 40) by the corporate profit-oriented media — television, radio, newspapers, magazines and film — cultivate the dominant ideology, mold mindless ways of understanding;,and homogenize everyday life experiences. The mass media play a significant role in transforming the individual into a compliant consumer. There is little escape from images of 'appropriateness' with which communities are relentlessly bombarded. These images become increasingly real as we develop relationships between ourselves and the items we see, hear and read. In fact, we buy the images; we believe in the necessity of that which is produced by the media. In fact, we emulate the images and engage in mindless and mimetic gestures of conformity as we become what we purchase. The central political and economic role of mass culture in manipulating through propaganda and thought control destroys reason, literacy and imagination. The "society of the spectacle" distracts, stupefies and paralyzes the public conscience. Instead of political action, the public is captured by the text of spectator sports, advertising and other mass cultural forms. For Chomsky (1989), a careful scrutiny of the subtext reveals that illusions are necessary in order to maximize the profits of corporate interests. In Europe, TV ads alone are expected to double to $36 billion by the end of this decade. The Pacific Rim's newly emerging TV market is already reaching $14 billion (Lippman, 1992: A21). Consciousness is constricted, advocacy muted and inequalities legitimized within these cultural codes. The media generally exploit and manipulate in the interests of power and self-preservation. The media convey socially sanctioned messages about the consequences of unacceptable conduct.

Television newscasts, for example, are especially effective in capturing our attention and then hold us 'captive' to "edited" accounts of "reality" slotted between commercials. The audience "gets close to the action," believing it knows immediately what is happening through the cultural intervention of a newscaster, impeccably attired to conjure up further images of integrity and professionalism . The audience is consumed in identifying with the messenger's so-called "objective" accounts, liturgical drama or self-evident truths. Ultimately, the audience is watching itself and the reproduction of dominant values without any discrimination of the misinformation. Through commercials, the economically powerful corporations hit hard in demonstrating that their consumerism will improve the quality of life for the audience. The audience buys the message and is programd to believe that the purchase of more and more products enhances success, sociability and physical attractiveness — to name only a few benefits. Newscasters cannot simply be viewed as representing their own personal and organizational interests; rather they are also cultural workers reproducing dominant ideologies. Cultural managers share class interests and associations with state and business managers and other privileged sectors (Chomsky, 1992: 93). To serve the interests of the powerful, the media must present a tolerably realistic picture of the world by sacrificing professional integrity and honesty. But the media are only one part of a larger doctrinal mirror; other parts include journals of opinion, the schools and universities, academic scholarship and so on. This doctrinal system diverts the attention of the unwashed masses and reinforces basic social values: passivity, submissiveness to authority, the overriding virtue of greed and personal gain, lack of concern for others, fear of real or imagined enemies, etc (*ibid.*, 95). The goal is to keep the bewildered masses even more bewildered, if not incarcerated. Therefore, it is unnecessary and culturally counter-productive for the 'audience' to become bothered with what is happening in the world. In fact, if people see too much of a disturbing reality, they may think about changing their roles as spectators (*ibid.*, 95). Simply, the bewildered herd is supposed to follow the recipe orders and enjoy the fast food meals neatly prepared and seductively packaged by corporate interests.

Just as Emperor Nero fed the Roman citizens a loaf of bread and a circus to placate them (Visano, 1985), the world is exposed to Cable Network News (CNN), a culturally American enterprise, as a global TV news channel now available in 137 countries. In general, television programs are a major US export, worth $2.3 billion annually. There are more than one billion television sets world-wide, and world-wide spending for television programming is $65 billion and growing ten percent annually (Lippman, 1992: A21).

Television programs are deliberately addictive, reflecting romanticized notions of escape under the guise of entertainment. The French, for example, spend more time watching TV than working (*ibid.*). Likewise, the film industry promotes conservative ideologies regarding the "survival of the slickest," laissez-faire vigilante-ism, technological marvels, fantasies etc. This co-optation of popular values diverts attention away from socio-political struggles inherent in an unequal society. Resistance and oppositional projects are simply dismissed as deviant or dangerous and thereby unprofitable.

Music too creates passivity. Whatever gets aired on the radio has been carefully appraised by radio stations attentive to their respective corporate sponsors and advertisers. The mainstream aired music is not creative, critical or confrontational. Rather, the criterion of conformity triumphs; a 'perfect fit' or conceptual congruence between profit and consumption is met. For instance, only a select representation of rap music that is deemed to be functionally and financially beneficial will enjoy official recognition.

The media allude insidiously to a general consensus in order to gloss over class struggles. All communication is filtered through a language that uses an innocuous tenor, hypnotic tone and vacuous themes that proclaim the benefits of corporate capital.

Hegemony, Consent and Antonio Gramsci (1891–1937)

To fully appreciate the importance of culture in social relations we now turn to the sophisticated investigations of Antonio Gramsci who forged an extremely innovative understanding of ideology as a constitutive dimension of structure and agency. In the *Quaderni del Carcere* [Notebooks] (1971) Gramsci accredited cultural facts, cultural activities and cultural fronts as necessary alongside the economic and the political. He attributed considerable significance to facts of culture and thought in the development of history. Culture complements force; ideological struggles are historical driving forces. That is, culture is a site of domination and resistance. Gramsci added:

> Capitalist private property dissolves every relation of common interest, binds and confuses conscience.... The life of men, the gains of civilization, the present, the future are in perpetual danger (1971: 31).

A key to the improvement of society was the transformation of thinking, the liberating potential of culture, the expression of creative capacities and the formulation of a new class order. For example, one's will and consciousness play fundamental roles in changing attitudes. This logic provides a sustained critique of economic determinism (Wolff, 1989), characteristic of a particular dogmatic Marxism. Gramsci's heterodoxy restores a sense of subjectivity and places significance on the intervention of human actors, not brute economic forces, as the primary movers of history (Nteta, 1987). Class rule in advanced capitalism is based on ideological-cultural hegemony within a civil society (Pontusson, 1980). Gramsci provided a theory of political action in which a more popular form of Marxism is possible.

Let us now focus more fully on Gramsci's inquiries into *hegemony* — a concept that is central to any criminalizing process. Answers to the following questions — how is crime constructed, what explains the cohesive nature of the criminal label and why do those criminalized, especially the oppressed, accept passively these routine disparaging designations — to name only a few may be found in the Gramscian concept of hegemony. In 1926, Gramsci re-formulated the dynamic concept of hegemony. Hegemony is the key element of political thought (Luksic, 1989), a condition that secures the consent of the exploited. Hegemony is a sophisticated carceral process incorporated by the ruling elite to silence opposition through accommodation (serving some public interest) and containment (absorbing opposition). Hegemony, as a flexible process and as a deeply entrenched structure only benefits the ruling class (Benney, 1983) by institutionalizing values and goals. Hegemony is the equilibrium between leadership based on consent and domination based on coercion. It is a form of cultural, moral and ideological rule based on force and consent over subordinated groups. In examining hegemony, the interests and tendencies of the subordinated groups need to be studied (Gramsci, 1988: 211).

Hegemony is constructed, renewed and re-enacted through a complex series or processes of struggle (Hall, 1988: 54). Gramsci identified procedures such as elections, collective bargaining and the courts in resolving conflict. The concept of hegemony is leadership based on the *consent* of the ruled, a consent secured by the diffusion and popularization of ruling class views. Consent of the ruled to their ongoing exploitation flows from the capitalists' hegemonic practices situated in all institutions of the state and civil society. Hegemony, achieved through institutions of civil society, is the predominance of one class over other classes through consent rather than force. This consent is manifested through a generic loyalty to the ruling class by virtue of their position in society. This position also warrants the ruling class to uphold the prevailing traditions and mores of the period. Historically, the ruling class developed this hegemony through a level of homogeneity, self-consciousness and organization articulated within economic, ethical, social, philosophic and political frames. Hegemony meant the permeation, throughout society, of an entire system of values, attitudes, morality, or beliefs that is supportive of controlling class interests. This prevailing consciousness is internalized and becomes part of a 'common sense.' Mouffe (1988: 103), in discussing the Gramscian "hegemonic principle" as the articulation of demands coming from different groups, explores the two ways in which demands can be advanced. First, there is "hegemony by neutralization" — a process wherein demands are taken into account to resolve antagonisms without transforming society. Second, there is "expansive hegemony," a process that links demands with other struggles to establish a chain of equivalence while respecting the autonomy and specificity of the demands of different groups. Stated differently, hegemony is an order in which dominant lifestyles and versions of reality are diffused throughout the society in all institutional and private manifestations, informing with its spirit all tastes, morality, customs, religious and political principles and all social relations, particularly in their intellectual and moral connotations (Merrington, 1968: 21). For Gramsci, hegemony cannot be purely ideological, since it must have as its foundation the domination of a particular social bloc in economic activity (Hall, 1988: 54). Hegemony fulfils a role that brute coercion could never perform especially in justifying deprivation and encouraging passivity. The dominant value system and its integrative effects penetrate everyday practices — all aspects of the social order. Gramsci cautioned, however:

> ...though hegemony is ethico-political, it must also be economic, must necessarily be based on the decisive function exercised by the leading group in the decisive nucleus of economic activity (1988: 211-212).

Gramsci went beyond Marx in defining *hegemony* as a political and cultural predominance of the working class and its party aimed at securing the compliance of other groups. Hegemony, the rule of consent rather than force, is the legitimation of revolution by a comprehensive culture (*ibid.*). Gramsci also articulated the hegemony of a proletariat, a consciousness of its own identity. Hegemonies rise and fall in their abilities to solve problems of conflict.

Ideology

But why do people and communities "consent" to their own exploitation? Why, for example, do people generally acquiesce to harmful economic, legal and social policies and applications? Simply, and on a more interactional level, how do we explain the deference to control especially by those individuals or groups who have been singled out or designated as dangerous?

The process of imposing definitions of crime is directly related to the maintenance of hegemony. Crime signifies disturbance, a challenge to hegemony and a clash with authorities. By subverting structural relations and resisting ideological integration, crime is oppositional. Dominant power blocs struggle to maintain domination by absorbing crime and rendering it useful without disrupting the status quo. Whenever institutions of authority perceive real or symbolic threats, the political dangers of crime are immediately constructed. Ideology as a constellation of ideas is manifested in all aspects of social life and performs powerful policing functions. As a cultural form, ideology legitimates social control. Ideology directs cognitions, evaluations and ideals with its links to political economy and the state. Ideology encapsulates by distorting material conditions and privileges. For example, the dominant ideology incorporates and displays features from other ideologies. For instance, liberty and freedom have become powerful instruments of domination. From challenges made under the Charter of Rights and Freedoms to the invasion of foreign sovereign states, the ideology of freedom has been appropriated conveniently. It is interesting to observe how the Canadian state selectively applies this rationale in providing support to the "reform movements" in the Soviet Union or Eastern Europe in the early 1990s and not to the struggles of our own Aboriginal people.

Ideology is a signifying system that manipulates and creates new discourses. But a dominant discourse must be expressed in the lived experiences of subjects in order to establish consent through common sense understandings. Accordingly, it is imperative that common sense be established to serve state and economic bloc interests. Ideologies therefore, are discourses flowing through subjects who use them to construct their self-identities. Common sense, as Hall (1988: 55) clarifies, is itself a structure of popular ideology, a spontaneous conception of the world, thoughts sedimented into everyday reasoning. Ideology becomes part of a generalized 'knowledge.' The dominant sectors not only transmit that knowledge which will legitimize their interests, but also deprive the dominated sectors of routine access to other versions of the 'truth.' The dissemination of everyday information is sufficient to satisfy elementary levels of popular curiosity.

Again, ideological control, for Gramsci, does not rest solely with repressive forces of the State, but rather on the ability of the dominant culture to secure compliance and inhibit any class consciousness or control over moral values. This ideological hegemony of the ruling classes is received by the masses as common sense, which blinds them to their own actual experiences (Counihan, 1986). Belief becomes intellectualized as ideology affects social behavior.

Ideology functions cognitively as a mode of self-interpretation. Ideology overwhelms in its persuasion rather than in its prescriptions. Resistance is difficult since this complex normative belief system, according to Gramsci, is both an independent and a related factor in the maintenance of state power. As Szymansky (1978: 176) notes:

> [F]ew people ever challenge the fundamental assumptions of capital-

ism and the state structure. People usually accept the basic "rules of the game" and judge the system to be fundamentally just and legitimate even if they see many specific failings. Such a high level of voluntary acceptance of the system does not occur spontaneously. The state must work very hard to produce such sentiments and reinforce them once they are created. The capitalist state maintains the legitimacy of the capitalist state through propagating procapitalist values and attitudes in schools.... This positive mechanism is supplemented by repressive measures designed to check the spread of anticapitalist and anti-state consciousness.

It is naive, however, to assume that ideology is simply an external conspiracy of repressive institutions of control. Following Gramsci and Foucault, one could argue that the dominant ideology cultivates helplessness by acting more locally through the consenting and docile individual. That is, ideology is inscribed as the social control of common-sense. Consequently, ideology contributes to the construction of consciousness. Ideological hegemony creates and homogenizes a self that enhances opportunities for control. An intriguing complement to Gramsci's "consent" is Foucault's (1979) "docile body." The latter is "a body that may be subjected, used, transformed and improved" (*ibid.*: 136). The individual is a reality manufactured by institutions of power (*ibid.*: 194). Ideology enhances a panoptic form of control; the construction of perceptions which guide action. As Foucault suggests:

> We are in the society of the teacher-judge, the doctor-judge, the social-worker judge; it is on them that the universal reign of the normative is based. This carceral network has the greatest support, in modern society, of the normalizing of power (1979: 304).

As Habermas (1974) admonishes, the meanings and symbols of the dominant ideology prevent critical thinking by penetrating social processes, language and individual consciousness. Ideology transforms the self into a subject; individuals adopt versions of the truth for themselves. But, as a sophisticated means of domination, ideology also succeeds in creating processes of self-subordination. The actor represses, deprives and denies self-autonomy by projecting a billiard-ball or assembly-line conception of self. As Friedenberg writes:

> Even a dog, Mr Justice Brandeis once observed, knows the difference between being stumbled over and being kicked. But an underdog,

apparently, can be persuaded to deny there is a difference, or at least to bite another underdog rather than the men who are kicking him (1975: xvii).

The vulnerability and credulity of the individual is promoted in the dominant liberal ideology which forever talks about the language of individuality — a disciplined or conformist subjectivity.

The dominant ideology monopolizes the means of mental production. Within post-structuralist interpretations, the objective dimension of self — the "me"— consists of mass-mediated images, and the "I" arises through the process of meaning deconstruction and reconstruction, that is, signifying practices (Kristiva, 1976). The self is a productive entity occupying a space between dominant discourses. The dominant discourse in the popular culture, however, is consistent with official conceptions.

Of considerable importance in contemporary Western societies are ideologies regarding private property, individual rights, family values, and "law and order" crusades which justify and buttress positions of privilege. Ideologies are not only filters through which we are encouraged to make sense of everyday routines but are also conceptual canopies under which control is constructed. A failure to mimic those values rewarded in the dominant culture invites serious repercussions. Liberalism highlights private/public distinctions and avoids any analyses of the political aspects of institutions outside of government (Adamson, 1987).

But this personal prison or incarcerated self is fundamentally related to legal practices. Structures of conformity are constituted through discursive disciplines. Juridic narratives are designed to reproduce control, to confer object-like realities to crime. Privilege channels thought within conformist boundaries (Chomsky, 1989: vii). As a mode of discourse, the patho-centric script of crime objectifies challenges by invoking existing codes. Legal, scientific and professional discourses dominate the language of control. As professionals adhere to a privatized, mystifying and techno-bureaucratic language, a cornucopia of meaningless jargon and mythologies emerge. Communication is convoluted, meanings are degraded, debates remain limited, and rituals that exalt privilege are rewarded. Moreover, specialized vocabularies conceal the "services" rendered to the dangerous classes and justify further interventions. For example, the defendant is transformed as a dependent, victims become witnesses, political prisoners are treated as common criminals. Numerous linguistic devices — rhetoric, jargon, cliches, or commonsensical 'sayings' — provide the appearance or the form that glosses over content and meaning.

Language, a central feature of all cultures, is both ideological and illusory. Language commoditizes control by structuring dependency relations. The process and structure of language as well as lexicons and vocabularies are anchored in particular histories that circumscribe acceptable expressions and marginalize differences. In other words, both the spoken and written language reflect limitations. Only when the actor surrenders to a codified and lethargic language is he or she considered to be both 'in' and 'of' society. Even the everyday chatter of familiar banter fails to move beyond the cultural script. Language objectifies, stultifies and disciplines expressions of self-awareness. The regularization of language is culturally necessary if talk is to be meaningful within this corporate landscape. These normative dimensions assign privilege to stereotypic language enabling the impoverishment of vocabularies and the domination of the technical, the efficient and the objective. Language presupposes a context of rules which cannot be contradicted (Gellner, 1959: 56). For Barthes (1973), language contributes to myth-making, language politicizes myths by claiming reference to signifying signs, and the clarity of language is misleading. Language is a performance that renders the more articulate as more legitimate and hence more knowledgeable. How then is it possible to transcend this cultural code? What facilitates the articulation of a critical consciousness that moves beyond the servitude of ideology, seductive illusions or self-paralysis?

Towards a Critical Cultural Approach

A program of cultural incarceration prevents social emancipation. Gramsci elaborated:

> The demagogy, the trickery, the untruth, the corruption of capitalist society are not accidental by-products of its structure; they are inherent in the disorder, in the unleashing of brutal passions, in the ferocious competition in which and by which capitalist society lives (1971: 31).

The "social" in our being is constituted in the context of cultural control. An awareness of one's own interpretive framework as part of a hegemonic force will lead to higher forms of self-consciousness. For Gramsci, there is no abstract 'human nature,' fixed and immutable, but instead there exists human nature as the totality of historically determined social relations (1971, 1985, 1988). That is, for Gramsci a program of mass education based on the aspirations, rights and duties of the subordinated groups untied to the monopoly of the bourgeoisie is imperative. A renewed emphasis on humanism encourages the development of a new consciousness. The masses had to educate themselves and gain independence from the bias of those who own the organs of public opinion in order to transform and authenticate their lived realities. Education as acculturation enhances an awareness of objective conditions that need to be mastered (Gramsci, 1971: 34). An emphasis on good sense replaces common sense (*ibid.*, 328).

Critical self-consciousness signifies the creation of intellectual cadres. Intellectuals provide a political approach that is liberating and not accommodating to the dominant culture, that is, these change agents are transformative. The intellectual as a key figure in transforming cultural hegemony must communicate meaningfully with workers thereby freeing each others' intellectual capacities. The development of a new consciousness that highlights the role of ideas in changing cultural phenomena in keeping with proletarian ideals is consistent with the reflective capacity of historical subjects. Lamentably, traditional intellectuals have given expression to the legitimacy of the status quo. As mercenaries, apologists, gaolers or as puppets, they advance knowledge that serves the capitalist mode of production; their ideas remain far removed from the people they encounter especially as they continue to market a certain culture. Moreover, Gramsci located the ideational control of intellectuals on vertical and horizontal axes. On the horizontal axis, there are traditional intellectuals, the learned members — artists, philosophers, writers who characterize the historical continuity in their work as incremental knowledge. On the other extreme, there are organic intellectuals, who are directly related to the economic structures and perform economic tasks, control and adopt technology. Organic intellectuals are also positioned vertically ranging from co-opted technicians at the bottom to those at the top — the elite, who amass fortunes and dominate all others. In North America, oppositional cultural politics among academics is sporadic, if not negligible. The orthodoxies of conventional texts, the rigidity of disciplinary boundaries, careerism, the dysfunctional nature of bureaucratic learning centers, the dearth of financial incentives as well as the constant backdrop of state repression tend to dismiss, trivialize and punish prospective contestants who have developed an appreciation of the importance of oppositional pedagogy. Conventional texts resist (Kumar, 1990) any commitment to the concrete world that articulates a need for social change. A philosophy of praxis, however, establishes a new language and sense of community. Constitutive unities were, in themselves, a product and process of politics (Golding, 1992: 45).

Intellectual activity is related to moral reform. The intellectual is not a priest or prophet; the challenge for the intellectual is to encourage the best for the disadvantaged such as the working class. For Gramsci, the transformation of the dominant cultural hegemony is both an inevitable process and a prerequisite for the creation of a political consciousness, a class action that turns hegemony against itself. The dissemination of ideas is an important role for the intellectual who is skilfully equipped to articulate extant injustices. The driving force for the intellectual, according to Gramsci, lies in the ability to transform the dominant bourgeois hegemony through the propagation of a new world view, to facilitate the creation of a new consciousness and to harness energy in the service of the proletariat. Gramsci was concerned in creating a new being, not in bettering one's material lot at the expense of others — a new way of perceiving the universe and one's position relative to it. For Gramsci, the party was fundamentally significant in effecting change. He notes: "intellectual and moral reform has to be linked with a program of economic reform — indeed the program of economic reform is precisely the concrete form in which every intellectual and moral reform presents itself" (1971: x).

How then is authenticity possible in light of such hegemonic forces? Modern ideologies may be confronted by creating a new core of intellectuals and by educating the masses of people. Marxism, for Gramsci, is a practical tool in stimulating and motivating people. The need to build a new intellectual and moral order arises concretely. Cultural praxis is a mode of resistance that transcends the reality of domination. Under certain circumstances people collectively undermine cultural constraints which prevent the full realization of individual and collective potential. Ethical priority assigns significance of both praxis over theory, and action over principle. A moral epistemology is of great utility in fostering intellectual and moral reform (Carceres, 1990).

Critical pedagogy and praxis are transformative in abandoning the metanarrative in favor of a will to power. The professorate, for instance, is urged to become activist in avoiding assimilation, conquest and the service of the elites (Ginsburg, 1987). True intellectuals are actively engaging with the masses in formulating a worldview that reflects the experience of the disadvantaged and not to merely perpetuate the status quo. For Gramsci, all humans are intellectuals. Traditionally, intellectuals have been socialized within the dominant classes and therefore have served the interests of the powerful. Organic intellectuals are those generated from the working class and represent a danger to the powerful. They are dangerous because they act as well as think critically (Bodenheimer, 1976). Through the development of a counter hegemony — that is, the hegemony of the proletariat — an alliance of different social groups is forged. Mass energies need to be enlisted in the struggle for ideological hegemony and in the construction of a progressive popular community out of the cleavages of the old society. Neither a leaderless mass movement nor the elite-run party can successfully forge a socialist identity that universalizes different struggles for narrow interests. During the initial period of struggle a war of position rages in which the role of the party is to lead the 'cultural-ideological' battle for moral-intellectual development. During military confrontations, the centralized combat party of professional revolutionaries assume primary importance in the" war of movement" (Boggs, 1972). Revolution as a war of position is extremely relevant because it conceives change not as mere seizure of power but as the building of a new culture of counter-hegemony into an historical bloc which over time becomes the state (Adamson, 1978). A war of "position" (led by the party) and of "manoeuvre" (Hall, 1986) by the hegemonic party, as conceived in the works of Gramsci, must lead the nation as a whole and may have to sacrifice the short-term interests of its own social class in order to attain its broader goals (Garner and Garner, 1981).

From Gramsci we learn to move beyond simple and banal re-assertions that power is conditioned by diverse ownership configurations. Rather, opportunities exist and await a critical recognition and interrogation of the modalities of power. The incremental transformation of radical research and the concretization of complex issues certainly inspire an authentic commitment to social theory. We learn that we are well-equipped to grapple with the problems inherent in uniting theory and practice. Knowledge of current trends in theorizing and research is gradually becoming more susceptible to wider dissemination and not strictly limited to academic activities. By removing certain institutional blinders which hinder social inquiry, we are witnessing the embrace of a critical imagination. Subsequently, this exercise includes the more innovative elaboration of conventional sets of instruments, techniques of analysis as well as levels of empirical applications.

Criminological theory and research, for example, are social enterprises which are historically grounded and environmentally influenced. To one observing the social aspects of crime research, one is puzzled by the relative paucity of historiography. An understanding of the concrete context of social inquiry has not been consistently or ardently pursued in terms of the obvious links between sociology and history. Crudely stated, the task of theorizing is to avoid exclusivity and to encourage the synthetic links between structural, dynamic and substantive categories-relations, process and the actor, setting or product, respectively. For instance, thought, class and political economy involve interrelated operationalized concomitants. The study of crime, with its sole emphasis on the imputations of meanings, risks the danger of becoming a barren and prosaic chronicle of exotic or even erotic inventories. It must be conceded that the central theme in crime is not the idea of designation but instead the combination of subjective activities, accomplishments and the processes which have stimulated the meanings of these products. Admittedly, the center of activities are located in the theorist-researcher who must be recognized as mediating a collective historical existence. That is, in constructing theories, this actor is a driving force articulating applications or empirically demonstrating conditions and contradictions in social relations. The product of this labor is a unification of existence and being. As empirical observers, we are equipped with an instructive dialectical methodology for unravelling the inter-relatedness of concepts, techniques and applications. This presupposes not only an attitude of contemplation alone but also discovery, presentation and praxis. Far too often, we routinely extrapolate or factor single variables as predominant according to a programmatic teleology, and underscore the immense significa-tion of conscience, action and the intricacies of our own contingencies. The dynamics of change, for example, require the testing of hypotheses in practice.

A theory of control acquires conviction in reference to historical and political aspirations. This inquiry into deviance is unavoidably sensitized by notions of struggle. The centrality of inequality indicates that despite rhetorical claims of the state's unyielding neutrality, crime is a negotiable cultural commodity politicized and de-politicized by state officials. Likewise, although crime is depicted as remotely related to the economic order, the economic order figures prominently in the options available to the subjected and in the lingering concerns of those subjecting. Attempts to reconstruct a humanist-liberationist philosophy of crime research warrant a compelling integration of real conditions and possibilities, those productive forces and relations which structure class, control and conflict. Research is not a closed process, rectilinear or dialectical but rather an open ongoing negotiation which cannot be separated easily from powerful influences.

A central liability to the development of a theoretical analysis of the concept of law and the practice of criminalization is the liberal assumption about society. The liberal state is dependent upon notions of individualism, the relationship of the individual to the state. Individual liberty, therefore, becomes abstracted from social relations.

A critical cultural critique rejects liberal assumptions about equality, artificial claims about individual merit and the benefits of competition. Liberalism, with its emphasis on egoism and self-fulfilment, must be juxtaposed against the history of human inequality (Lakoff, 1964: 196). Equality, it is submitted, is a collective and historical accomplishment. Liberal discourses repeatedly avoid equality and accessibility 'to' resources while stressing the individual's relative freedom of action. As Marx (1956) writes, we are not abstractions but rather we are the totality of social relations. In other words, we are all connected to historical socio-political trends. Any authentic emergence of the social, therefore, invites a commitment to advocacy, education, reform and coalition-building. Although the separation of research and praxis is a theoretical abstraction, this division is not without consequences. Despite its pejorative or deviant designation, "praxis" gives impetus to intellectual confrontations, reduces uncertainty and highlights contradictions. As reiterated, research is a social undertaking which assists in the development of consciousness of one's interactions and labor. A vital strength of this type of activity is the creation and use of information. The tendency to pay lip-service and emasculate a praxis model reduces research as a means to an end and becomes derivative of yet another ideological reproduction and legitimation. Notwithstanding the invisible controls of funding sources, crime research can construct some noteworthy inquiries about and excursions into social reform. A radical analysis suggests that the problematic of linkage cannot be resolved without examining the fundamental distribution of resources in our society. The analyst interrogates the role of the state and how it serves to enrich certain interests by bestowing disproportionate benefits upon them or helping to preserve the structure of class inequality. A critical approach is sensitive to what state agencies profess to be doing, exposing contradictions and recommending remedies. Ultimately, the social responsibility of the theorist is to probe into fundamental social values upon which law and state depend. By developing a method of thinking, a perspective which strives to clarify the mythologies and ambiguities, crime research calls for a review of these ideals.

Lamentably, a further limitation which impacts on the pursuit of viable social directions is the moral posturing within the intellectual community. At one end, radical scholars are maligned and readily dismissed for using crime as a political platform. At the other end of the spectrum, indictments flourish against researchers who reinforce extant control practices and perpetuate the existing order. In any event, this divisive morality play trivializes actual and potential contributions. Suffice it to say that there are those intellectuals who function organically within economic tasks in exercising technical capacities. There are also those intellectuals for whom historical continuity is expressed in their work as cumulative knowledge (Cammett, 1967: 20).

Transformation is required not only in the direction of crime research but also in the attitudes towards this investigation. Upon further reflection, we must move beyond the convenience of traditional texts and clarify the significance of culture, political economy and the state to crime theories and methods. Our analysis of culture demonstrates the primacy of lived experience that is often overlooked. The lived experience is marked by inequalities throughout the social landscape. Traditional approaches reduce crime to a language that echoes prevailing ideologies, thereby reducing deviance to what Foucault describes as the "chatter" of criminology. Crime is about inequality and the attendant social relations in local sites that are culturally contextualized. Identity, as a constituted subject, is the negotiated outcome of this process.

The Social Self and the State of Culture: Prospects or Problems?

Within the dominant culture crime exists as an ideology, an immutable and objective reality independent of processes of social definition. This paper, however, suggests that the rules designating crime are created, injected with meaningfulness and dependent upon interpretations. Rules are constituted by social actors, internalized, applied and accepted as legitimate. The leading or dominant ideologies, as accounting processes of deviance and control, are emergent articulations of moral justifications. Identities or even truths about offenders are advanced as 'total,' exhaustive catch-all clauses, rather than as cultural indicators that situate ideology. Ideology is an interpretive scheme for making sense of phenomena, a methodology for classifying the social world. The available arsenal of moral rhetoric, language and symbols facilitate the phenomenal reduction of consciousness and intention. Juxtaposed against powerful ideological antidotes, there is the problem of discovering material conditions and developing an emancipatory consciousness. Emancipation requires a struggle for equal rights and duties, the abolition of class rule supported by the privileged, white, heterosexual male culture that shapes relations.

Traditional texts reduce crime to monolithic, totalized and essentialist conceptions within a narrow theoretical vision and an overarching conceptual closure. As Miles (1982: 3) notes, we should question "the way in which common sense discourse has come to structure and determine academic discourses...." Crime, as everyday practices and policies articulating hegemonic processes, is both contradictory and complementary; crime resists as well as accommodates. But, how are these antagonistic positions made possible? The situated reality of crime/conformity and the wider political economy is mediated by culture. Cultural reproduction mediates forms and processes. Crime is constituted by various, albeit conflicting, discourses. Although as Laclau and Mouffe (1985) argue, the social is never totalized or complete, the "social" is mystified. The social aspects of crime are constituted within hegemonic processes that become concretized at local, cultural and political levels. What therefore passes typically as a subject of crime or punishment is well filtered in a manner that fully ignores racism, sexism, homophobia, and class inequalities. Accordingly, it is incumbent upon students of socio-legal studies to move beyond the text and interrogate the subtext of their own credulity and appreciate the fragmented nature of their own subjectivity. It is certainly an onerous task to move beyond cultural roles and expectations incubated within resources like law, bureaucratic rationality, media, manufactured common sense, education employment etc. Such fundamental issues of race, gender, homophobia and class continue to escape scrutiny. Not only are we compelled to inquire into these battlegrounds, we focus on the cultural reworkings of larger discourses that deviantize non-whites, women and the poor. Avoiding the reductivist tendencies inherent in the dominant culture, we argue that race, gender, homophobia and class are interconnected and should be analyzed as social constructions within the context of reproducing oppression. From Antonio Gramsci we learn about the social contexts within which people develop ideological commitments. Ideology is a superstructure understood in a specific historical context through concepts of hegemony, civil society, the state, the party and intellectuals (Hall, Lumley and McLennan, 1977). Gramsci's wider vision of the state incorporates coercion and hegemony. With the latter, the state is able to achieve consensus even among the working class through ideological means, of which the most important is nationalism or patriotism. Likewise, welfare measures have been valuable in controlling the working class (Cueva and Martinez-Baracs, 1980). Cultural contradictions exist in the juxtapositions of freedom, liberty and equality. A culture industry has constructed crime as an attribute of individualism. The individual person or group who challenges is thereby rendered insignificant. As a text, culture is an expression of ideology which should be examined in terms of its production and consumption, its practices of encoding and decoding. A method to clarify the coding processes is available in a critical cultural theory, a long overdue perspective in the study of crime. Critical cultural approaches provide an alternative framework in appreciating the form and content of crime. Implicitly, this social theory captures the historical development of ideologies, the role of the popular culture, and the communications of privilege. Awareness is culturally mediated and linguistically conditioned. How can we deconstruct codes of thinking and critically elaborate on the conditions and consequences of deviance? The answer rests with consciousness, knowing ourselves and our location (Gramsci, 1971; Jouve, 1991: 8). Is self-knowledge an illusion? Is it not created in "social" relationships which link the conveniently

bifurcated private and public worlds? Crime has become a discursively constituted institution that patterns and reproduces agency. Unfortunately, this
concept is seldom explicated or understood in relation to a cultural debate,
molded by illusions. Instead, conformity is invested with the attribution of the
real, the functional and stable. The awareness of being different and seeing
differences inspire manoeuvres that banish or correct the crises of challenges to
the texts. Differences are fundamentally linked to individual identities which are
reproduced as showpieces, as escapes or imported foreign values.

Culture privileges rich, white, able, heterosexual males while crime incarcerates the disadvantaged who are always pacified and transformed as objects
of exchange and constituted as convenient commodities. Human subjects, from
the criminal subject to the subjected reader are controlled by the censorial gaze
of institutions. Culture mediates, provides rationality and articulates a language
that magically defines conformity by imposing a familiar context and generates
universal commitment. Those, not belonging are "out of place" according to the
strictures of culture. As Wiseman (1989: 42) suggests, the "suspicion is that
culture and tradition have transmuted the institutionally chosen into the naturally
given." This emphasis on cultural and social, however, does not deny the equally
significant forces of economic and political structures.

Control is ideologically situated within institutions that shape self-identity.
These protective cocoons filter trouble, funnel interpretations and marginalize
differences according to universalized framed experiences and inoculated
reflexivities (Giddens, 1992: 3). A priori conditioning incites intolerance. In this
paper the common-sense knowledge and beliefs about conformity are
problematized. Culture manipulates sanctions by defining disturbances as local
accommodations to contests or as totalizing narratives of trouble that warrant
closure, containment and coercion. Culture transforms subjects into social
objects, "demonizes" differences or deviantizes the devalued "other." This
essentially cultural subject is not analyzed as the totalizing consequence of
structure nor the simply reduced situated agent. Rather, this construction of
control reflects a dialectic of consent framed within power relations and logically
a resistance of the knowing subject. Culture is pluralistic, a panoply of discourses
and competing values. The "deviant" subject is segmented and articulated
within a variety of discourses that succeed in abstracting, mystifying and
decontextualizing social control. The reproduction of crime is a cultural practice,
a way of experiencing how the world persists, preserves and universalizes
inequality. Notwithstanding the banter of liberalism, culture ignores differences
and denies authenticity. The authentic — especially the concrete resistance or
the risks of one's own situation, is replaced by forms of pseudo-community

communication.

Culture is a barrier to communication that has become institutionalized within absolutist and corporate interpretive frameworks. Culture is ideologically constituted in the interests of state and corporate capital. Culture is the fabric that clothes and protects the body of privilege. Common culture is a myth. The commitment to many cultures disguises the subversion of racism, sexism, homophobia and class elitism. The text of a collective conscience is a moral binding force or the cement of a nation that articulates support for differences or pluralism within a "common" culture. As Homi Bhabha (1992: 235) notes:

> The dangers inherent in the concept of a contemporaneous "common" culture are not limited to politically conservative discourses. There is a pervasive, even persuasive, presence of such a paradigm in the popular rhetoric of multiculturalism.... lip service is paid to the representation of the marginalized. A traditional rhetoric of cultural authenticity is produced on behalf of the" common culture" from the very mouths of the minorities.

In addition, Bhabha adds: "The common culture promises 'individual' emancipation.... The common culture is the ideological purveyor of this 'impersonal' order of things" (1992: 242). Culture historicizes selectively the experiences of exclusion suffered by women, people of color and the poor.

Any project committed to cultural analysis interrogates the character of social order, locates this interrogation inside and outside the bounds of conventional rules, stepping back and looking into dominant institutional forms that marginalize those deemed to be the other — the criminal, the deviant. Culture, it is argued, "normalizes" exercises of power by constructing binary oppositions. Accordingly, our cultural approach relocates that which has occupied the margins to the center, decenters and de-essentializes the subject by highlighting how the subject is constructed in contradictions. The deference to "essential" elements of criminality is re-positioned and negotiable — always involving an open process of transformation. Culture has an extensive capacity to adapt to perturbations, conflict and resistance.

Various forms of cultural stretch, continuous and episodic, trigger passivity; the connective tissues of ideology allow the control fibers considerable elasticity or range of motion.

This dominant culture is replete with illusions which enhance alienation. As Marx advised: "Your very ideas are but the outgrowth of your bourgeois production and bourgeois property" (Tucker, 1978: 487). The conditions of the material environment deny creativity and restrict choices. This culture is material; culture surrenders to the material. Transformation of material productive forces leads to cultural changes. As Hook (1933: 41) explains:

> Marx realized that every culture is a structurally interrelated whole, and that any institutional activity, say religion or law, can be understood only in relation to a whole complex of other social activities.

In our society, the dominant culture is expressed through material consumption. Materialism is not simply the product of socialization but more importantly, an imperative of the economic order. The conditions of our materialism constitute concretized forms of existence. Materialism, as an ubiquitous feature of our culture, speaks on our behalf; that is, our sense of self emerges from material dependence. It is used as a code, to guide and classify coherently the ways in which we display our experiences, imagination and consciousness. As a condition, materialism subjugates, enslaves and binds. Much effort is invested in conforming to the ways we appear, what we possess, fixated through materialism, resulting in self-enclosure. Identities and self-worth are carved by an enslaving materialism. As a society, we invest much in carving out conformity as measured according to our possessions. Fixation becomes self-enclosure. Materialism overwhelms as it mediates mental, emotional and social expressions. Although people living in advanced capitalist societies are not forced to be material, they consent and agree to be acquisitive. Materialism is personal to the extent that the dominant culture celebrates possessive individualism. Materialism manipulates especially as actors are expected to have an insatiable appetite for material gains. Those who do not have great material goods feel responsible for their sense of failure. Simultaneously, materialism is individualized-assessed in reference to individual efforts. The sanctity of the marketplace symbolizes advanced capitalist societies. Within series of exchanges in the marketplace of social relations, people treat each other as objects with prospective material value. Society is atomized as people are respected as individualized, independent, self-reliant — as "winners or losers" competing for more material. Wholesale consumerism objectifies the individual who is shaped according to material results of production and reproduction. Individuals are transformed into objects allied to products. Relations of production quantify the individual.

The socio-cultural processes occur within environments that assign currency to reproductive practices. Whether in the world of state propaganda, the projects of profit or the limits of knowledge, culture situates and is situated according to differences. The text of exclusive cultural enterprises succeeds in deviantizing outsiders by limiting the discourse to certain appropriate modes of thinking. A culture of impoverished ideas triumphs conceptual imperialism distorts, alienates and coerces into compliance an already susceptible general consciousness. Dominant modes of mental production divide, invade and colonize according to fundamental imperatives of capitalism. Differences that challenge or ideas that "really make a difference" are tortured into submission. Equally, ideas are celebrated, once calibrated and re-routed successfully in reference to a market-driven logic of materialism. Intrusions are not only state-sponsored but ideologically legitimated. Interestingly, embedded at all tiers of domination and unequal relations are cultural mythologies (law, justice and morality), cultural colonizers ("the experts") and cultured consent (vulgar co-optation and generalized manufactured consensus) that succeed in further marginalizing the "disadvantaged." Notwithstanding the roles played by the law and criminal justice system in protecting 'individual' rights and the rights to 'property,' major cultural institutions like religion, family and education silence the helpless and dispossessed through a dubious benevolence of dependency.

The state hegemonizes by responding to challenges with repression and terror. The state tolerates trouble as long as it can be absorbed into existing political-economic structures. Culture provides the 'reason,' the *modus operandi*, by setting up a series of non-thinking values, developing subtleties of crime as pain or pleasure, advancing mythologies regarding justice and morality, creating panics which restore standards of conformity. Typically, the unsuspecting general public does not think about various ways of thinking. Instead, shared symbols are ritually obeyed. Norms are inherited for generations without any critical debates. Schematically, Figure 1 describes the dynamic relationships that structure crime.

Figure 1

Political Economy *   <<<>>> State **   <<<>>> Ideology ***

---

*	= class interests, privilege, production, property/profit, ideology, state and history.
**	= law, coercion, history, bureaucracy, effectiveness, criminal "justice," ideology and political economy.
***	= culture, morality/"justice," violence, materialism, inequalities (race, gender, class, sexual orientation), popular images, law, political economy, state and history.

References

Adamson, W. 1987. "Gramsci and the Politics of Civil Society," *Praxis-International*. 7(Oct-Jan.) 3-4: 320-339.

Adorno,T. and M. Horkheimer. 1989. Selections from "The Culture Industry: Enlightenment as Mass Deception" in Gottlieb, R. (ed.), *An Anthology of Western Marxism*. NY: Oxford University Press.

Barthes, R. 1975. *The Pleasure of the Text*. Translated by R. Miller. NY: Hill and Wang.

Bhabha, H. 1992. "A Good Judge of Character: Men, Metaphors and The Common Culture" in Morrison, T. (ed.), *Race-ing Justice, En-gendering Power*. NY: Pantheon.

Benney, M. 1983. "Gramsci on Law, Morality and Power," *International Journal of the Sociology of Law*. 11 (May) 2: 191-208.

Bodenheimer, T. 1976. "The Role of Intellectuals in Class Struggle," *Synthesis*. 1 (Summer) 1: 20-27.

Boggs, C. 1972. "Gramsci's Prison Notebooks: Part 2," *Socialist Revolution*. 2 (Nov.-Dec.) 6: 29-56.

Cammett, J. 1967. *Antonio Gramsci and the Origins of Italian Communism*. Stanford: Stanford University.

Carceres, M. 1988. "Gramsci, Religion and Socio-economic Systems," *Social Compass*. 35 (2-3): 279-296.

Chomsky, N. 1987. *Pirates and Emperors*. New York: Black Rose.

Chomsky, N. 1989. *Necessary Illusions*. Toronto: CBC Enterprises.

Chomsky, N. 1992. *What Uncle Sam Really Wants*. Berkeley: Odonian Press.

Counihan, C. 1986. "Antonio Gramsci and Social Science," *Dialectical Anthropology*. 11 (1): 3-9.

Cueva, A. and A. Martinez-Baracs. 1980. "An Interview with Christine Buci-Glucksman," *Revista Mexicana de Sociologia*. 42 (Jan- Mar) 1: 289-301

Foucault, M. 1965. *Madness and Civilization: A History of Insanity in the Age of Reason*. New York: Vintage.

Foucault, M. 1970. *The Order of Things*. New York: Vintage.

Foucault, M. 1979. *Discipline and Power*. N.Y.:Pantheon.

Foucault, 1977b. "The Political Function of the Intellectual," *Radical Philosophy*. 17: 12-14.

Foucault, M. 1980. *Power and Knowledge: Selected Interviews and Other Writings 1972-1977*. Edited by C. Gordon. New York: Vintage.

Foucault, M. 1980b. *The History of Sexuality. Vol I*. New York: Vintage.

Friedenberg, E. 1975. *The Disposal of Liberty and Other Industrial Wastes*. Garden City: Doubleday.

Garner, L. and R.Garner. 1981. "Problems of the Hegemonic Party: The PCI and the Structural Limits of Reform," *Science and Society*. 45 (Fall) 3: 257-273.

Gellner, E. 1959. *Words and Things*. London: Routledge and Kegan.

Giddens, A. 1992. *Modernity and Self-Identity*. Stanford: Stanford University Press.

Ginsburg, M. 1987. "Contardictions in the Role of Professor as Activist," *Sociological Focus*. 20 (April) 2: 11-122.

Golding, S. 1992. *Gramsci's Democratic Theory: Contributions to Post-Liberal Democracy*. Toronto: University of Toronto Press.

Gramsci, A. 1957. *The Modern Prince and Other Writings*. Edited By L. Marks. New York: International Publishers.

Gramsci, A. 1971. *Prison Notebooks: Selections*. Translated by Q. Hoare and G. Smith. New York: International Publishers.

Gramsci, A. 1985. *Antonio Gramsci: Selections From Cultural Writings*. Edited by D. and G. Nowell-Smith. Translated by W. Melhower. Cambridge: Harvard University Press.

Gramsci, A. 1988. *Antonio Gramsci: Selected Writings 1916-1935*. Edited by D. Forgacs. New York: Schocken.

Hall, S. 1986. "Gramsci's Relevance for the Study of Race and Ethnicity," *Journal of Communication Inquiry*. 10 (Summer) 2: 5-27.

Hall, S. 1988. "Toad in the Garden: Thatcherian Among the Theorist" in Nelson, C. and L. Grossberg (eds.), *Marxism and the Interpretation of Culture*. Urbana: University of Illinois.

Hall, S., B. Lumley and G. McLennan. 1977. "Politics and Ideology: Gramsci," *Working Papers in Cultural Studies*. 10: 45-76.

Hook, S. 1933. *Towards the Understanding of Karl Marx: A Revolutionary Interpretation*. Ann Arbor: University of Michigan.

Itwaru, A. 1989a. *Mass Deception*. Toronto:Terebi.

Itwaru, A. 1989b. *Critiques of Power*. Toronto:Terebi.

Jouve, N.W. 1991. *White Woman Speaks With Forked Tongue: Criticism as Autobiography*. London: Routledge.

Key, W. B. 1981. *Subliminal Seduction*. NY: Signet.

Kristiva, J. 1976. Signifying Practice and Mode of Production," *Edinburgh Magazine*. (1).

Kumar, A. 1990. "Towards Postmodern Marxist Theory: Ideology, State and the Politics of Critique," *Rethinking Marxism*. 3: (Fall-Winter) 3-4: 149-155.

Laclau, E. and C. Mouffe .1985. *Hegemony and Socialist Strategy: Towards a Radical Democratic*. New York: New Left Books.

Lakoff, S. 1964. *Equality in Political Philosophy*. Boston: Beacon.

Lippman, J. 21 December 1992, p. A21. "How Television Is Shaping World's Culture,"

Toronto Star.

Lippman, W. 1992b. *Public Opinion.* New York: Macmillan.

Luksic, I. 1989. "The First Uses of the Concept of Hegemony in Gramsci," *Athropos.* 20 (3-4): 348-354.

Marx, K. 1956. *Selected Writings in Sociology and Social Philosophy.* Edited by T. Bottomore. London: Watts and Company.

Marx, K. 1965. *Capital. Volume 1: A Critical Analysis of Capitalist Production.* Moscow: Progress Publishers.

Marx, K. 1965. *Selected Correspondence.* Moscow: Progress Publishers.

Marx, K. 1969. *Capital: A Critique of Political Economy.* Moscow: Progress Publishers.

Merrington, J. 1968. "Theory and Practice in Gramsci's Marxism," *Sociologist Register.* 145-176.

Miles, R. 1982. *Racism and Migrant Labor.* London: Routledge and Kegan Paul.

Mouffe, C. 1988. "Hegemony and New Political Subjects: Toward a New Concept of Democracy" in Nelson, C. and L. Grossberg (eds.), *Marxism and the Interpretation of Culture.* Urbana: University of Illinois.

Nteta,C. 1987. "Revolutionary Self Consciousness as an Objective Force Within the Process of Liberation: Biko and Gramsci," *Radical America.* 21 (Sept.- Oct.) 5: 55-61.

Pontusson, J. 1980. "Gramsci and Euro-communism: A Comparative Analyses of Class Rule and Socialist Transition," *Berkeley Journal of Sociology.* 24-25: 185-248.

Syzmanski, A. 1978. *The Capitalist State and the Politics of Class.* Cambridge: Cambridge University Press.

Tucker, R. (ed.). 1978. *The Marx-Engels Reader.* Second Edition. W.W. Norton and Company.

Visano, L.A. 1985. "Crime, Law and the State" in Fleming, T. (ed.), *The New Criminologies in Canada.* Toronto: Oxford University Press.

Visano, L.A. Forthcoming. *Beyond The Text.* Toronto: HBJ-Holt.

Wiseman, M.B. 1989. *The Ecstasies of Roland Barthes.* London: Routledge.

Wolff,R. 1989. "Gramsci, Marxism and Philosophy," *Rethinking Marxism.* 2 (Summer) 2: 41-54.

Appendix

An Act Respecting Public and Reformatory Prisons

SHORT TITLE

Short title

1. This Act may be cited as the *Prisons and Reformatories Act*. R.S., c. P-21, s. l.

INTERPRETATION

Definitions
"lieutenant governor"

"Minister"

"prison"

"prisoner"

2. (1) In this Act,

"lieutenant governor" means lieutenant governor in council;

"Minister" means the Solicitor General of Canada;

"prison" means a place of confinement other than a penitentiary as defined in the *Penitentiary Act;*

"prisoner" means a person, other than

(a) a child within the meaning of the *Juvenile Delinquents Act,* chapter J-3 of the Revised Statutes of Canada, 1970, as it read immediately prior to April 2,1984, with respect to whom no order pursuant to section 9 of that Act has been made, or

(b) a young person within the meaning of the *Young Offenders Act* with respect to whom no order pursuant to section 16 of that Act has been made, who is confined in a prison pursuant to a sentence for an offence under an Act of Parliament or any regulations made thereunder.

Custody

(2) Where a prisoner is temporarily outside a prison but under the direct charge, control or supervision of an officer or employee of a prison, the prisoner is in custody for the purposes of this Act and any other Act of Parliament. R.S., 1985, c. P-20, s. 2; R.S., 1985, c. 35 (2nd Supp.), s. 29.

COMMITTAL, RECEPTION AND TRANSFER OF PRISONERS

Warrant of committal

3. Where a person is sentenced or committed to imprisonment in a prison, it is sufficient compliance with the law, notwithstanding anything in the *Criminal Code,* if the warrant of committal states that the person was sentenced or committed to imprisonment in a prison for the term in question, without stating the name of any particular prison. R.S., c. P-21, s. 3; 1976-77, c. 53, s. 45.

4. (1) [Repealed, R.S., 1985, c. 35 (2nd Supp.), s. 30]

Transfers between provinces

(2) The governments of the provinces may enter into agreements with one another providing for the transfer of prisoners from a prison in one province to a prison in another province.

Effect of transfer

(3) A prisoner transferred under an agreement made pursuant to subsection (2) shall be deemed to be lawfully confined in the receiving prison and is subject to all the statutes, regulations and rules applicable in the receiving prison. R.S., 1985, c. P-20, s. 4; R.S., 1985, c. 35 (2nd Supp.), s. 30.

Transfers from penitentiaries to prisons

5. (1) The Minister may, with the approval of the Governor in Council, enter into an agreement with the government of any province for the transfer of inmates from any penitentiary in Canada to any prison in that province.

Idem

(2) The Commissioner of Corrections or a member of the Canadian Penitentiary Service designated by the Commissioner may direct transfers of inmates in accordance with agreements entered into under subsection (1).

Effect of transfer

(3) An inmate transferred under this section or under an agreement made pursuant to any other lawful authority shall be deemed to be lawfully confined in the receiving prison and is subject to all the statutes, regulations and rules applicable in the receiving prison. R.S., 1985, c. P-20, s. 5; R.S., 1985, c. 35 (2nd Supp.), s. 31.

EARNED REMISSION

Remission

6. (1) Every prisoner serving a sentence shall be credited with fifteen days of remission of the sentence in respect of each month and with a number of days calculated on a pro rata basis in respect of each incomplete month during which the prisoner has applied himself industriously, as determined in accordance with any regulations made by the lieutenant governor of the province in which the prisoner is imprisoned, to the program of the place of confinement in which the prisoner is imprisoned.

Computing remission

(2) The first credit of earned remission credits pursuant to subsection (1) shall be made not later than the end of the month next following the month the prisoner is received into a prison and thereafter a credit of earned remission shall be made at intervals of not more than three months.

Idem

(3) Where a prisoner was received into a prison before July 1, 1978, the date of the first credit of earned remission referred to in subsection (2) is August 31, 1978 and the subsequent intervals run from that date.

Forfeiture

(4) Every prisoner who, having been credited with earned remission, commits any breach of the prison rules is, at the discretion of the person who determines that the breach has been committed, liable to forfeit, in whole or in part, the earned remission that stands to the credit of the prisoner and that accrued to the prisoner after July 1, 1978. R.S., 1985, c. P-20, s. 6; R.S., 1985, c. 35 (2nd Supp.), s. 32.

TEMPORARY ABSENCE

Temporary Absence

7. (1) Where, in the opinion of an officer designated by the lieutenant governor of the province in which a prisoner is confined, it is necessary or desirable that a prisoner should be absent, with or without escort, for medical or humanitarian reasons or to assist in the rehabilitation of the prisoner, subject to subsection (2) the absence may be authorized by that officer for an unlimited period for medical reasons and for a period not exceeding fifteen days for humanitarian reasons or to assist in the rehabilitation of the prisoner.

Approval of provincial parole board

(2) Where, in a province, a provincial parole board has been appointed pursuant to section 12 of the *Parole Act*, the lieutenant governor of the province may order that no absence without escort be authorized except under that authority and with the approval of the provincial parole board. R.S., c. P-21, s. 8; 1976-77, c. 53, s. 45.

Effect on date of release

8. Where a prisoner is granted a temporary absence and the date on which he is due to be released falls within the period of that temporary absence, the prisoner shall, for the purposes of all entitlements accruing to him on release, be deemed to have been released on the day on which the temporary absence commenced. R.S., c. P-21, S. 9; 1976-77, c. 53, s. 45.

YOUNG OFFENDERS

9. [Repealed, R.S., 1985, c. 24 (2nd Supp.), s. 49]
10. and
11. [Repealed, R.S., 1985, c. 35 (2nd Supp.), s. 33]
12. Repealed, R.S., 1985, c. 1 (1st Supp.), s. 1]
13. [Repealed, R.S., 1985, c. 1 (1st Supp.), s. 2]

Canadian Bill of Rights

An Act for the Recognition and Protection of Human Rights and Fundamental Freedoms.

8-9 Elizabeth II, c. 44 (Canada)

Preamble

[Assented to 10th August 1960]

The Parliament of Canada, affirming that the Canadian Nation is founded upon principles that acknowledge the supremacy of God, the dignity and worth of the human person and the position of the family in a society of free men and free institutions;

Affirming also that men and institutions remain free only when freedom is founded upon respect for moral and spiritual values and the rule of law;

And being desirous of enshrining these principles and the human rights and fundamental freedoms derived from them, in a Bill of Rights which shall reflect the respect of Parliament for its constitutional authority and which shall ensure the protection of these rights and freedoms in Canada:

Therefore her Majesty, by and with the advice and consent of the Senate and House of Commons of Canada, enacts as follows:

PART 1
Bill of Rights

Recognition and declaration of rights and freedoms

1. It is hereby recognized and declared that in Canada there have existed and shall continue to exist without discrimination by reason of race, national origin, colour, religion or sex, the following human rights and fundamental freedoms, namely,

(a) the right of the individual to life, liberty, security of the person and enjoyment of property, and the right not to be deprived thereof except by due process of law;

(b) the right of the individual to equality before the law and the protection of the law;

(c)   freedom of religion;
(d)   freedom of speech;
(e)   freedom of assembly and association; and
(f)   freedom of the press.

Construction of law   2.   Every law of Canada shall, unless it is expressly declared by an Act of Parliament of Canada that it shall operate notwithstanding the *Canadian Bill of Rights,* be so construed and applied as not to abrogate, abridge or infringe or to authorize the abrogation, abridgment or infringement of any of the rights or freedoms herein recognized and declared, and in particular, no law of Canada shall be construed or applied so as to

(a)   authorize or effect the arbitrary detention, imprisonment or exile of any person;

(b)   impose or authorize the imposition of cruel and unusual treatment or punishment;

(c)   deprive a person who has been arrested or detained
(i)   of the right to be informed promptly of the reason for his arrest or detention,
(ii)   of the right to retain and instruct counsel without delay, or
(iii)   of the remedy by way of *habeas corpus* for the determination of the validity of his detention and for his release if the detention is not lawful;

(d)   authorize a court, tribunal, commission, board or other authority to compel a person to give evidence if he is denied counsel, protection against self-criminalization or other constitutional safeguards;

(e)   deprive a person of the right to a fair hearing in accordance with the principles of fundamental justice for the determination of his rights and obligations;

(f) deprive a person charged with a criminal offense of the right to be presumed innocent until proved guilty according to law in a fair and public hearing by an independent and impartial tribunal, or of the right to reasonable bail without just cause; or

(g) deprive a person of the right to the assistance of an interpreter in any proceedings in which he is involved or in which he is a party or a witness, before a court, commission, board or other tribunal, if he does not understand or speak the language in which such proceedings are conducted.

Duties of Minister of Justice

3. The Minister of Justice shall, in accordance with such regulations as may be prescribed by the Governor in Council, examine every proposed regulation submitted in draft form to the Clerk of the Privy Council pursuant to the *Regulations Act* and every Bill introduced in or presented to the House of Commons, in order to ascertain whether any of the provisions thereof are inconsistent with the purposes and provisions of this Part and he shall report any such inconsistency to the House of Commons at the first convenient opportunity.

Short title

4. The provisions of this Part shall be known as the *Canadian Bill of Rights*.

PART II

Savings

5. (1) Nothing in Part I shall be construed to abrogate or abridge any human right or fundamental freedom not enumerated therein that may have existed in Canada at the commencement of this Act.

"Law of Canada" defined

(2) The expression "law of Canada" in Part I means an Act of the Parliament of Canada enacted before or after the coming into force of this Act, any order, rule or regulation thereunder, and any law in force in Canada or in any part of Canada at the commencement of this Act that is subject to be repealed, abolished or altered by the Parliament of Canada.

Jurisdiction of Parliament

(3) The provisions of Part I shall be construed as extending only to matters coming within the legislative authority of the Parliament of Canada.

Constitution Act, 1982
Schedule B
Part I

Canadian Charter of Rights and Freedoms

Whereas Canada is founded upon principles that recognize the supremacy of God and the rule of law:

Guarantee of Rights and Freedoms

Rights and freedoms in Canada
1.   The *Canadian Charter of Rights and Freedoms* guarantees the rights and freedoms set in it subject only to such reasonable limits prescribed by law as can be demonstrably justified in a free and democratic society.

Fundamental Freedoms

Fundamental freedoms
2.   Everyone has the following fundamental freedoms:
          (a)   freedom of conscience and religion;
          (b)   freedom of thought, belief, opinion and expression, including freedom of the press and other media of communication;
          (c)   freedom of peaceful assembly; and
          (d)   freedom of association.

Democratic Rights

Democratic rights of citizens
3.   Every citizen of Canada has the right to vote in an election of members of the House of Commons or of a legislative assembly and to be qualified for membership therein.

Maximum duration of legislative bodies
4.   (1)   No House of Commons and no legislative assembly shall continue for longer than five years from the date fixed for the return of the writs at a general election of its members.

Continuation in special circumstances
          (2)   In time of real or apprehended war, invasion or insurrection, a House of Commons may be continued by Parliament and a legislative assembly may be continued by the legislature beyond five years if such continuation is not opposed by the votes of more than one-third of the members of the House of Commons or the legislative assembly, as the case may be.

Annual sitting of legislative bodies
5.   There shall be a sitting of Parliament and of each legislature at least once every twelve months.

Mobility Rights

Mobility of citizens

6.   (1)   Every citizen of Canada has the right to enter, remain in and leave Canada.

Right to move and gain livelihood

(2)   Every citizen of Canada and every person who has the status of a permanent resident of Canada has the right

(a)   to move and take up residence in any province; and

(b)   to pursue the gaining of a livelihood in any province.

Limitation

(3)   The rights specified in subsection (2) are subject to

(a)   any laws or practices of general application in force in a province other than those that discriminate among persons primarily on the basis of province of present or previous residence; and

(b)   any laws providing for reasonable residency requirements as a qualification for the receipt of publicly provided social services.

Affirmative action programs

(4)   Subsections (2) and (3) do not preclude any law, program or activity that has as its object the amelioration in a province of conditions of individuals in that province who are socially or economically disadvantaged if the rate of employment in that province is below the rate of employment in Canada.

Legal Rights

Life, liberty and security of person

7.   Everyone has the right to life, liberty and security of the person and the right not to be deprived thereof except in accordance with the principles of fundamental justice.

Search or seizure

8.   Everyone has the right to be secure against unreasonable search or seizure.

Detention or imprisonment

9.   Everyone has the right not to be arbitrarily detained or imprisoned.

Arrest or detention

10.   Everyone has the right on arrest or detention

(a)   to be informed promptly of the reasons therefor;

(b)   to retain and instruct counsel without delay and to be informed of that right; and

(c)   to have the validity of the detention determined by

way of *habeas corpus* and to be released if the
detention is not lawful.

Proceedings in criminal and penal matters
11.   Any person charges with an offence has the right

(a)   to be informed without unreasonable delay of the
specific offence;

(b)   to be tried within a reasonable time;

(c)   not to be compelled to be a witness in proceedings
against that person in respect of the offence;

(d)   to be presumed innocent until proven guilty
according to law in a fair and public hearing by an
independent and impartial tribunal;

(e)   not to be denied reasonable bail without just cause;

(f)   except in the case of an offence under military law
tried before a military tribunal, to the benefit of trial
by jury where the maximum punishment for the
offence is imprisonment for five years or a more
severe punishment;

(g)   not to be found guilty on account of any act or
omission unless, at the time of the act or omission, it
constituted an offence under Canadian or
international law or was criminal according to the
general principles of law recognized by the
community of nations;

(h)   if finally acquitted of the offence, not to be tried for
it again and, if finally found guilty and punished for
the offence, not to be tried or punished for it again;
and

(i)   if found guilty of the offence and if the punishment
for the offence has been varied between the time of
commission and the time of sentencing, to the
benefit of the lesser punishment.

Treatment or punishment
12.   Everyone has the right not to be subjected to any cruel and unusual
treatment or punishment.

Self-crimination
13.   A witness who testified in any proceedings has the right not to have
any incriminating evidence so given used to incriminate that witness in
any other proceedings, except in a prosecution for perjury or for the
giving of contradictory evidence.

Interpreter
14.   A party or witness in any proceedings who does not understand or
speak the language in which the proceedings are conducted or who
is deaf has the right to the assistance of an interpreter.

Equality Rights

Equality before and under law and equal protection and benefit of law
15.   (1)   Every individual is equal before and under the law and has the right to the equal protection and equal benefit of the law without discrimination and, in particular, without discrimination based on race, national or ethnic origin, colour, religion, sex, age or mental or physical disability.

Affirmative action programs
       (2)   Subsection (1) does not preclude any law, program or activity that has as its object the amelioration of conditions of disadvantaged individuals or groups including those that are disadvantaged because of race, national or ethnic origin, colour, religion, sex, age or mental or physical disability.

Official Languages of Canada

Official languages of Canada
16.   (1)   English and French are the official languages of Canada and have equality of status and equal rights and privileges as to their use in all institutions of the Parliament and government of Canada.

Official languages of New Brunswick
       (2)   English and French are the official languages of New Brunswick and have equality of status and equal rights and privileges as to their use in all institutions of the legislature and governments of New Brunswick.

Advancement of status and use
       (3)   Nothing in this Charter limits the authority of Parliament or a legislature to advance the equality of status or use of English and French.

Proceedings of Parliament
17.   (1)   Everyone has the right to use English or French in any debates and other proceedings of Parliament.

Proceedings of New Brunswick legislature
       (2)   Everyone has the right to use English or French in any debates and other proceedings of the legislature of New Brunswick.

Parliamentary statutes and records
18.   (1)   The statutes, records and journals of Parliament shall be printed and published in English and French and both language versions are equally authoritative.

New Brunswick statutes and records
(2)   The statutes, records and journals of the legislature of New Brunswick shall be printed and published in English and French and both language versions are equally authoritative.

Proceedings in courts established by Parliament
19.   (1)   Either English or French may be used by any person in, or in any pleading in or process issuing from, any court established by Parliament.

Proceedings in New Brunswick courts
(2)   Either English or French may be used by any person in, or in any pleading in or process issuing from, any court of New Brunswick.

Communications by public with federal institutions
20.   (1)   Any member of the public in Canada has the right to communicate with, and to receive available services from, any head or central office of an institution of the Parliament or government of Canada in English or French, and has the same right with respect to any other office of any such institution where
(a)   there is a significant demand for communications with and services from that office in such language; or
(b)   due to the nature of the office, it is reasonable that communications with and services from that office be available in both English and French.

Communications by public with New Brunswick institutions
(2)   Any member of the public in New Brunswick has the right to communicate with, and to receive available services from, any office of an institution of the legislature or government of New Brunswick in English or French.

Continuation of existing constitutional provisions
21.   Nothing in sections 16 to 20 abrogates or derogates from any right, privilege or obligation with respect to the English and French languages, or either of them, that exists or is continued by virtue of any other provision of the Constitution of Canada.

Rights and privileges preserved
22.   Nothing in sections 16 to 20 abrogates or derogates from any legal or customary right or privilege acquired or enjoyed either before or after the coming into force of this Charter with respect to any language that is not English or French.

Minority Language Educational Rights

Language of instruction
23.   (1)   Citizens of Canada
            (a)   whose first language learned and still understood is
                    that of the English or French linguistic minority
                    population of the province in which they reside, or
            (b)   who have received their primary school instruction
                    in Canada in English or French and reside in a
                    province where the language in which they
                    received that instruction is the language of the
                    English or French linguistic minority population of
                    the province, have the right to have their children
                    receive primary and secondary school instruction in
                    that language in that province.

Continuity of language instruction
        (2)   Citizens of Canada of whom any child has received or is
                receiving primary or secondary school instruction in English
                or French in Canada, have the right to have all their children
                receive primary and secondary school instruction in the same
                language.

Application where numbers warrant
        (3)   The right of citizens of Canada under subsections (1) and (2)
                to have their children receive primary and secondary school
                instruction in the language of the English or French linguistic
                minority population of a province
            (a)   applies wherever in the province the number of
                    children of citizens who have such a right is
                    sufficient to warrant the provision to them out of
                    public funds of minority language instruction; and
            (b)   includes, where the number of those children so
                    warrants, the right to have them receive that
                    instruction in minority language educational facilities
                    provided out of public funds.

Enforcement of guaranteed rights and freedoms
24.   (1)   Anyone whose rights or freedoms, as guaranteed by this
                Charter, have been infringed or denied may apply to a court
                of competent jurisdiction to obtain such remedy as the court
                considers appropriate and just in the circumstances.

Exclusion of evidence bringing administration of justice into disrepute
        (2)   Where, in proceedings under subsection (1), a court
                concludes that evidence was obtained in a manner that
                infringed or denied any rights or freedoms guaranteed by
                this Charter, the evidence shall be excluded if it is established
                that, having regard to all the circumstances, the admission of
                it in the proceedings would bring the administration of justice

into disrepute.

General

Aboriginal rights and freedoms not affected by Charter
25.　　The guarantee in this Charter of certain rights and freedoms shall not be construed so as to abrogate or derogate from, any aboriginal treaty or other rights or freedoms that pertain to the aboriginal peoples of Canada including
　　　　(a)　　any rights or freedoms that have been recognized by the Royal Proclamation of October 7, 1763; and
　　　　(b)　　any rights or freedoms that now exist by way of land claims agreements or may be so acquired.

Other rights and freedoms not affected by Charter
26.　　The guarantee in this Charter of certain rights and freedoms shall not be construed as denying the existence of any other rights or freedoms that exist in Canada.

Multicultural heritage
27.　　This Charter shall be interpreted in a manner consistent with the preservation and enhancement of the multicultural heritage of Canadians.

Rights guaranteed equal to both sexes
28.　　Notwithstanding anything in this Charter, the rights and freedoms referred to in it are guaranteed equally to male and female persons.

Rights respecting certain schools preserved.

An Act Respecting Penitentiaries

SHORT TITLE

Short Title

1.　This act may be cited as the *Penitentiary Act*. R.S., c. P-6, s. 1.

INTERPRETATION

Definitions
"Commissioner"

"inmate"

"Minister"

"penitentiary"

"Service"

2.　In this Act,
　　"Commissioner" means the Commissioner of Corrections referred to in section 5;
　　"inmate" means a person who, having been sentenced or committed to a penitentiary, has been received and accepted at a penitentiary pursuant to the sentence or committal and has not been lawfully discharged therefrom or from any other place pursuant to section 23.1;
　　"Minister" means the Solicitor General of Canada;
　　"penitentiary" means an institution or facility of any description, including all lands connected therewith, that is operated by the Service for the custody, treatment or training of persons sentenced or committed to penitentiary, and includes any place declared to be a penitentiary pursuant to subsection 3(1) or (2);
　　"Service" means the Correctional Service of Canada referred to in section 4. R.S., 1985, c. P-S, s. 2; R.S., 1985, c. 35 (2nd Supp.), s. 15.

PENITENTIARIES

Place declared a
penitentiary

Idem

3.　(1)　The Commissioner may, by order, declare any prison as defined in the *Prisons and Reformatories Act* or any hospital to be a penitentiary in respect of any person or class of persons.
　　(2)　The Governor in Council may, by order, declare any place to be a penitentiary.

Provincial approval required

(3) No prison, hospital or place administered or supervised under the authority of an Act of the legislature of a province may be declared a penitentiary under subsection (1) or (2) until an officer designated by the lieutenant governor of the province in which the prison, hospital or place is located gives his approval.

Extended meaning of penitentiary

(4) For the purposes of any law of Canada relating to escapes and rescues of prisoners, a penitentiary shall be deemed to include any place at or in which an inmate, prior to the inmate's lawful discharge from custody, is required by this Act or the regulations, or by an officer of the Service, to be or remain.

Lands constituting penitentiary

(5) In any proceedings before a court in Canada in which a question arises concerning the location or dimension of lands alleged to constitute a penitentiary, a certificate, purporting to be signed by the Commissioner, setting out the location or description of the lands as constituting a penitentiary, is admissible in evidence and in the absence of any evidence to the contrary is proof that the lands as located or described in the certificate constitute a penitentiary.

Custody

(6) Where an inmate is temporarily outside a penitentiary but under the direct charge, control or supervision of a member of the Service, the inmate is in custody for the purposes of this Act and any other Act of Parliament. R.S., 1985, c. P-5 s. 3; R.S., 1985, c. 35 (2nd Supp.), S. 16.

CORRECTIONAL SERVICE OF CANADA

Correctional Service of Canada

4. There shall continue to be a penitentiary service in and for Canada, to be known as the Correctional Service of Canada. R.S., 1985, c. P-5, s. 4; R.S., c. 35 (2nd Supp.), s. 17.

COMMISSIONER

Commissioner   5.

The Governor in Council may appoint an officer to be known as the Commissioner of Corrections who, under the direction of the Minister, has the control and management of the Service and all matters connected therewith. R.S., c. P-6, s. 4; 1976-77, c. 53, s. 36.

Administration of Parole Service   6.

The portion of the staff of the National Parole Board known as the National Parole Service shall be under the control and management of the Commissioner who, in addition to the duties described in section 5, is responsible, under the direction of the Minister, for the preparation of cases of parole and the supervision of inmates to whom parole has been granted or who have been released on mandatory supervision pursuant to the *Parole Act*. 1976-77, c. 53, s. 37.

OFFICERS AND EMPLOYEES

Directors   7.   (1)

The Minister may appoint officers of the Service to be known as Directors of Divisions and Regional Directors.

Maximum number   (2)

The maximum number of officers in each class and their salaries shall be as prescribed by the Treasury Board. R.S., c. P-6, s. 5.

Other officers and employees   8.   (1)

The Commissioner, under the direction of the Minister, may appoint such other officers and employees of the Service as are necessary for the administration of this Act, and, in respect of those appointments, the preferences provided in the *Public Service Employment Act* in respect of military service apply.

Ranks and grades   (2)

The ranks and grades of officers and employees appointed by the Commissioner under subsection (1), the maximum number of persons to be appointed to each of those ranks and grades and their salaries shall be as prescribed by the Treasury Board. R.S., c. P-6, s. 6.

Oath	9.	(1)	Every officer and employee of the Service shall, before entering on the duties of his office, take the oath of allegiance and, in the case of an officer, an oath of office in the following form: I, , solemnly swear that I will faithfully, diligently and impartially execute and perform the duties required of me as an officer of the Canadian Penitentiary Service and will well and truly obey and perform all lawful orders that I receive as such, without fear, favour or affection of or toward any person. So help me God.
Authority to administer		(2)	The oath prescribed by subsection (1) and any other oath or declaration that may be necessary or required may be taken by the Commissioner before any judge, magistrate or justice of the peace having jurisdiction in any part of Canada, and by any other officer of the Service before the Commissioner or any officer in charge of an institution or any person having authority to administer oaths or take or receive affidavits. R.S., c. P-6, s. 7.
Tenure	10.	(1)	Officers and employees of the Service hold office during pleasure.
Suspension		(2)	The Commissioner may, where the Commissioner considers it in the interests of the Service, suspend from duty any officer or employee of the Service.
Idem		(3)	The officer in charge of a penitentiary may, where that officer considers it in the interests of the Service, suspend from duty any officer or employee of the Service who is under his jurisdiction. R.S., c. P-6. s. 8.
Member of Service	10.1		All officers and employees of the Service shall be deemed to be members of the Service. R.S., 1985, c. 35 (2nd Supp.), s. 18.

When Commissioner and Deputy Commissioner absent

11.

In the event that the Commissioner and Deputy Commissioner are absent or unable to act or their offices are vacant, the senior Divisional Head at the headquarters of the Service has the control and management of the Service and all matters connected therewith, and for those purposes the senior Divisional Head may exercise all the powers of the Commissioner under this Act or any other Act of Parliament. R.S., c. P-6, s. 9.

Peace officer

12.

The Commissioner may, in writing, designate any member of the Service or each member of a class of members of the Service to be a Peace officer and a member so designated is a peace officer in every part of Canada and has all the powers, authority, protection and privileges that a peace officer has by law. R.S., 1985, c. P-5, s. 12; R.S., c. 35 (2nd Supp.), s. 19.

HEADQUARTERS

Headquarters

13. (1)

The headquarters of the Service and the offices of the Commissioner shall be at Ottawa.

Regional headquarters

(2)

The Commissioner may establish regional headquarters of the Service and fix the location of regional offices. R.S., c. P-6, s. 11.

INVESTIGATIONS

Investigations

14.

The Commissioner may, from time to time, appoint a person to investigate and report on any matter affecting the operation of the Service and, for that purpose, the person so appointed has all of the powers of a commissioner appointed under Part II of the Inquiries Act, and section 10 of that Act applies, with such modifications as the circumstances require, in respect of investigations carried on under the authority of this section. R.S., c. P-6, s. 12.

COMMITTAL, RECEPTION AND TRANSFER OF INMATES

Warrant of committal	15.	(1)	Where a person is sentenced or committed to imprisonment for life, for an indeterminate period or for any term that is required to be served in a penitentiary, it is sufficient compliance with the law, notwithstanding anything in the Criminal Code, if the warrant of committal states that the person was sentenced for life, for an indeterminate period or for the term in question, as the case may be, without stating the name of any penitentiary to which the person is sentenced or committed.
Rules		(2)	The Commissioner may make rules naming the penitentiaries in which, in the first instance, persons sentenced or committed in any part of Canada to penitentiary shall be received.
Transfer		(3)	Where a person has been sentenced or committed to penitentiary, the Commissioner or any officer directed by the Commissioner may, by warrant under the hand of the Commissioner or that officer, direct that the person shall be committed or transferred to any penitentiary in Canada, whether or not that person has been received in the relevant penitentiary named in rules made under subsection (2).
		(4)	[Repealed, R.S., 1985, c. 35 (2nd Supp.), s. 20.
Custody in transit		(5)	A person shall be deemed to be in lawful custody anywhere in Canada if, (a) having been sentenced or committed to penitentiary, that person is in the custody of a person acting under the authority of the court that sentenced or committed him; or (b) having been directed to be transferred to another penitentiary, that person is in the custody of a person acting under the authority of the officer who directed the transfer. R.S., 1985, c. P-5, s. 15; R.S., 1985, c. 35 (2nd Supp.), s. 20.

Newfoundland	16.	(1)	Notwithstanding anything in this Act, every person who is sentenced by any court in Newfoundland to imprisonment for life, or for a term of years not less than two, shall be sentenced to imprisonment in the penitentiary operated by the Province of Newfoundland at the city of St. John's for the confinement of prisoners, and shall be subject to the statutes, rules, regulations and other laws pertaining to the management and control of that penitentiary.
Transfer from Newfoundland		(2)	Subsection 15(3) applies in respect of persons imprisoned under subsection (1), except that such a person shall not be transferred from the penitentiary mentioned in subsection (1) without the approval of an officer designated by the Lieutenant Governor of Newfoundland.
Agreement		(3)	The Minister may, with the approval of the Governor in Council, enter into an agreement with the Province of Newfoundland providing for the payment to the Province of the cost of maintaining the persons who are or have been sentenced or committed to penitentiary. R.S., 1985, c. P-5, s. 16; R.S., 1985, c. 35 (2nd Supp.), s. 21.

YUKON TERRITORY AND NORTHWEST TERRITORIES

Arrangements with provinces	17.	(1)	The Minister may, with the approval of the Governor in Council, arrange with the lieutenant governor of any province for the confinement, in the prisons or reformatories of that province, of persons convicted in the Yukon Territory or the Northwest Territories and for the compensation to be paid by the Government of Canada to the government of the province in respect of persons so confined.

Transfer

(2) Where an arrangement has been made under subsection (1), the Commissioner or any officer directed by the Commissioner may, by warrant under the hand of the Commissioner or that officer, direct the transfer of a person convicted in the Yukon Territory or the Northwest Territories to a prison or reformatory in a province in respect of which the arrangement applies, and the person shall, while being escorted to that prison, be deemed to be in a lawful custody.

Deeming

(3) A person who is confined in a prison or reformatory outside the Yukon Territory or the Northwest Territories pursuant to an arrangement made under subsection (1) shall, during the term of that person's sentence or period confined. R.S., c. P-6, s. 14.

SENTENCES OF LESS THAN TWO YEARS

Agreements with provinces

18. (1) The Minister, with the general or special approval of the Governor in Council, may on behalf of the Government of Canada enter into an agreement with the government of any province for the confinement in penitentiaries or any other institutions under the direction or supervision of the Service of persons sentenced or committed to imprisonment for less than two years for offences under this Act or any other Act of Parliament or any regulations made thereunder, but any such agreement shall include provisions whereby those persons shall be confined at the expense of the provincial government concerned.

Effect of confinement in penitentiary

(2) A person who is confined in a penitentiary or other institution pursuant to an agreement made under subsection (1) shall, during committal, be deemed to be lawfully confined and is subject to all the statutes, regulations, rules and orders applicable in the penitentiary or in the institution. R.S., c. P-6, s. 15; 1976-77, c. 53, s. 39.

RECEPTION OF INMATES

Time when persons may be received	19. (1)	A person who has been sentenced or committed to penitentiary shall not be received in a penitentiary until after the expiration of the time limited by law for an appeal, and thereupon that person may be received in a penitentiary whether or not that person has entered an appeal.
Election not to appeal	(2)	A person referred to in subsection (1) may, before the expiration of the time limited by law for an appeal, give written notice to the court that sentenced or committed that person that he elects not to appeal or abandons his appeal, as the case may be, and thereupon the time limited for appeal shall be deemed to have expired.
Transfer for preparation or presentation of appeal	(3)	Where the Commissioner or an officer of the Service designated by the Commissioner is satisfied that attendance of an inmate is required away from the penitentiary into which the inmate has been received, for the purpose of the preparation or presentation of an appeal from the inmate's conviction or sentence, the Commissioner or the officer designated by the Commissioner may issue a written direction to the officer in charge of the penitentiary into which the inmate has been received directing that officer, for that purpose, to transfer the inmate to a prison, common jail or other place, not being a penitentiary, in which persons who are charged with or convicted of offences are usually kept in custody. R.S., c. P-6. s. 16.
Medical certificate	20.	Subject to any relevant agreement that may be made under section 22, the officer in charge of a penitentiary is not required to accept a person into custody under a warrant of committal unless there is, in relation to that person, a certificate of a duly qualified medical practitioner that certifies that the person is free from dangerous, contagious or infectious disease. R.S., c. P-6, s. 17.

Custody before being
received into
penitentiary

21. (1) A person who by reason of subsection 19(1) is not received into a penitentiary or who by reason of section 20 is not accepted into custody shall be confined in any prison, common jail or other place, not being a penitentiary, in which persons who are charged with or convicted of offences are usually kept in custody.

Custody by keeper of
prison, common jail or
other place

(2) The keeper of any prison, common jail or other place referred to in other place subsection (1) or 19(3) to whom a person referred to in either of those subsections is delivered shall, on sufficient authority, receive, safely keep and detain that person under custody in the prison, common jail or other place until that person is returned to or transferred to a penitentiary or discharged from custody in accordance with law.

Sufficient authority

(3) The original of the warrant or other instrument by which a person referred to in subsection (1) or 19(3) is committed to or is to be imprisoned in a penitentiary, or a copy thereof duly certified by any judge or magistrate or by the clerk of the court in which that person was convicted, is sufficient authority for that person's detention in accordance with subsection (2). R.S., c. P-6, s. 18.

TEMPORARY ACCOMMODATION

Assisting person's
rehabilitation

21.1 Subject to the Commissioner's directives, where in the opinion of the member in charge of a penitentiary it would assist the rehabilitation of a person who has been released on parole or subject to mandatory supervision pursuant to the *Parole Act* to allow that person to be temporarily accommodated in that penitentiary, the member in charge may, on the request of the person, allow the person to stay temporarily in the penitentiary for such period of time and subject to such conditions as are specified by the member in accordance with the Commissioner's directives. R.S., 1985, c. 35 (2nd Supp.), s. 22.

TEMPORARY ACCOMMODATION

Mentally Ill Inmates   22. (1)   The Minister may, with the approval of the Governor in Council, enter into an agreement with the government of any province to provide for the custody, in a mental hospital or other appropriate institution operated by the province, of persons who, having been sentenced or committed to penitentiary, are found to be mentally ill or mentally defective during confinement in penitentiary.

Idem   (2)   Where no agreement has been made under subsection (1) between the Minister and the government of any province from which a mentally ill or mentally defective person is sentenced or committed to penitentiary, the officer in charge of the penitentiary may, on the advice of the penitentiary physician or psychiatrist, refuse to accept custody of that person under the sentence or committal or, if custody of that person has been accepted, may, under the authority of a written direction by the Commissioner, return that person to the prison or other place of confinement from which that person was received.

Diseased inmates   (3)   The Minister may, with the approval of the governor in Council, enter into an agreement with the government of any province to provide for the custody, in penitentiary hospitals, of persons who, having been sentenced or committed to a provincial prison, are found to be suffering from any dangerous, contagious or infectious disease during the sentence.

Deeming   (4)   A person who, pursuant to subsection (1), is confined in a provincial hospital or other institution shall, during the term of his confinement therein, be deemed to be confined in a penitentiary.

Idem   (5)   A person who, pursuant to subsection (3), is confined in a penitentiary hospital shall, during the term of his confinement therein, be deemed to be confined in a provincial prison. R.S., c. P-6, s. 19.

Discharge of diseased
inmates

23.

Where, on the day appointed for the lawful discharge of an inmate from a penitentiary, the inmate is found to be suffering from a disease that is dangerous, contagious or infectious, the inmate shall be detained in the penitentiary until such time as the officer in charge has made appropriate arrangements for the treatment of the inmate in an appropriate provincial institution or until the inmate is cured, whichever is the earlier. R.S., c. P-6, s. 20.

DISCHARGE OF INMATES GENERALLY

Place of discharge

23.1

An inmate may be discharged from a penitentiary or from any other place designated by the Commissioner's directives. R.S., 1985, c. 35 (2nd Supp.), s. 23.

Date of release

23.2 (1)

An inmate, other than a paroled inmate as defined in the *Parole Act,* who is entitled to be released shall be released during the daylight hours of the last working day prior to the ordinary release date of the inmate.

Definition of "working day"

(2)

For the purposes of subsection (1), "working day," in a province, means a day on which offices of the Public Service of Canada are generally open in that province. R.S., 1985, c. 35 (2nd Supp.), s. 23.

YOUNG INMATES

24.

[Repealed, R.S., 1985, c. 24 (2nd Supp.), s. 48]

EARNED REMISSION

Remission 25. (1) Subject to this section and section 26.1, every inmate shall be credited with fifteen days of remission of the sentence of the inmate in respect of each month and with a number of days calculated on a pro rata basis in respect of each incomplete month during which the inmate has been industrious, as determined in accordance with any Commissioner's directives made in that behalf, with regard to the program of the penitentiary in which the inmate is imprisoned.

Computing remission (2) The first credit of earned remission credits pursuant to subsection (1) shall be made not later than the end of the month next following the month the inmate is received into a penitentiary, and thereafter a credit of earned remission shall be made at intervals of not more than three months.

Idem (3) Where an inmate was received into a penitentiary before July 1, 1978, the date of the first credit of earned remission referred to in subsection (2) is August 31, 1978 and the subsequent intervals run from that date.

References to expiration of sentence according to law (4) For the purposes of this section and section 26.1, a reference to the expiration of a sentence of an inmate according to law shall be read as a reference to the day on which the sentence expires, without taking into consideration any remission standing to the credit of the inmate.

Effect of remission (5) An inmate is not entitled to be released from imprisonment, solely as a result of remission,

(a) prior to the expiration according to law of the sentence the inmate is serving at the time an order is made in respect of the inmate pursuant to paragraph 21.4(4)(a) of the *Parole Act,* as determined in accordance with section 20 of that Act at the time the order is made; or

(b) where the case of the inmate is referred to the Chairman of the National Parole Board pursuant to subsection 21.3(3) of the *Parole Act* during the six months immediately preceding the presumptive release date of the inmate, prior to the rendering of the decision of the Board in connection therewith.

Effect of direction not to be released as a result of remission

(6) Where an order is made in respect of an inmate pursuant to paragraph 21.4(4)(a) of the *Parole Act,* the inmate shall forfeit all statutory and earned remission standing to the credit of the inmate, whether accrued before or after the coming into force of this section.

Idem

(7) Any remission of sentence forfeited pursuant to subsection (6) shall not thereafter be recredited pursuant to subsection 24(3) of the *Parole Act.* R.S., 1985, c. P-5, s. 25; R.S., 1985,c. 34 (2nd Supp.). s. 10.

Forfeiture of earned remission

26. Every inmate who, having been credited with earned remission, is convicted in disciplinary court of any disciplinary offence is liable to forfeit, in whole or in part, the earned remission that stands to the credit of the inmate and that accrued after July 1, 1978 but no such forfeiture of more than thirty days shall be valid without the concurrence of the Commissioner or a member of the Service designated by the Commissioner, or of more than ninety days without the concurrence of the Minister. R.S., 1985, c. P-5, s. 26; R.S., 1985, c. 35 (2nd Supp.), s. 24.

No remission on revocation of mandatory supervision

26.1 (1) Where, following an order of the Board made pursuant to paragraph 21.4(4)(a) or (b) of the *Parole Act* or an order declaring that, at the time the case was referred to the Board, the inmate was serving a term of imprisonment that included a sentence imposed in respect of an offence mentioned in the schedule to the *Parole Act* that had been prosecuted by indictment and declaring that, in the opinion of the Board, the commission of the offence caused the death of or serious harm to another person, the inmate was released subject to mandatory supervision and the mandatory supervision is revoked, the inmate

(1) (a) shall, except in respect of a consecutive sentence or portion thereof imposed after the inmate's release subject to mandatory supervision and served prior to the revocation of the mandatory supervision, forfeit all statutory and earned remission standing to the credit of the inmate, whether accrued before or after the coming into force of this section; and

(b) is not entitled to be released from imprisonment, solely as a result of remission, prior to the expiration according to law of the sentence, as determined in accordance with section 20 of the *Parole Act,* that the inmate was serving on the date of release.

Idem

(2) Any remission of sentence forfeited pursuant to subsection (1) shall not thereafter be remitted or recredited pursuant to paragraph 25(2)(c) or (d) or subsection 25(3) of the *Parole Act.* R.S., 1985, c. 34 (2nd Supp.), s. 11.

PAROLE

27. [Repealed, R.S., 19485, c. 35 (2nd Supp.), s. 25.

TEMPORARY ABSENCE

Escorted temporary absence

28. Where, in the opinion of the Commissioner or the officer in charge of a penitentiary, it is necessary or desirable that an inmate should be absent, with escort, for medical or humanitarian reasons or to assist in the rehabilitation of the inmate, the absence may be authorized by

(a) the Commissioner, for an unlimited period for medical reasons and for a period not exceeding fifteen days for humanitarian reasons or to assist in the rehabilitation of the inmate; or

(b) the officer in charge, for a period not exceeding fifteen days for medical reasons and for a period not exceeding five days for humanitarian reasons or to assist in the rehabilitation of the inmate. R.S., c. P-6, s. 26; 1976-77, c. 53, s. 42.

Where inmate transferred to provincial institution

29. (1) Where, pursuant to an agreement made; under subsection 22(1), an inmate has been admitted to a provincially operated mental hospital or to any other provincially operated institution in which the liberty of the patients is normally subject to restriction, the officer in charge of the provincial institution may permit temporary absences with escort from that institution, within the limits prescribed in paragraph 28(b), when the officer is delegated that authority by the member in charge of the penitentiary in which the inmate was last confined.

Delegation of authority

(2) For the purposes of subsection (1), the member in charge of a penitentiary may delegate the authority to grant temporary absences to the officer in charge of the provincial institution described in subsection (1) either generally or for specific cases. R.S., 19485, c. P-5, s. 29; R.S., 19485, c. 35 (2nd Supp.), s. 26.

30. [Repealed, R.S., 1985, c. 35 (2nd Supp.), s. 26.

Effect on date of temporary release

31. Where an inmate is granted a temporary release absence and the day on which he is due to be released falls within the period of that temporary absence, the inmate shall, for the purpose of all entitlements accruing to him on release, be deemed to have been released on the day on which the temporary absence commenced. 1976-77, c. 53. s. 42.

PENITENTIARY INDUSTRY

Advisory Committee

32. (1) There shall be a committee called the Advisory Committee on Penitentiary Industry, to be appointed by the Minister and to consist of not more than nine persons chosen from the fields of industry, labour, government and the general public, to advise the Commissioner concerning industrial operations to be carried on by inmate labour in connection with the Service.

Expenses

(2) Members of the Advisory Committee appointed pursuant to subsection (1) shall serve without remuneration but are entitled to be paid reasonable travel and living expenses incurred by them while absent from their ordinary place of residence in connection with the work of the Committee. R.S., c. P-6, s. 27.

BUILDING AND WORKS

Commissioner's powers

33. The repair and maintenance of buildings and works in relation to penitentiaries and, to the extent specified by any order of the Governor in Council, the construction of those buildings and works, are under the control and direction of the Commissioner. R.S., c. P-6, s. 28.

COMPENSATION FOR DISABILITY OR DEATH

Minister may pay
compensation

34. (1) Subject to and in accordance with any regulations made under subsection 37(1), the Minister may pay compensation
   (a) to a discharged inmate for physical disability attributable to the inmate's participation in the normal program of a penitentiary; and
   (b) to the surviving spouse or dependent children of a discharged inmate or an inmate who died before the expiration of his sentence whose death is attributable to his participation in the normal program of a penitentiary.

Definition of
"discharged inmate"

(2) In this section, "discharged inmate" means an inmate who has been released as a result of the expiration of his sentence or the operation of remission or who has been released on parole other than day parole. 1976-77, c. 53, s. 43.

FORFEITURE OF CONTRABAND

Forfeiture

35. (1) Subject to subsections (2) and (3), where an inmate is convicted in disciplinary court of possession of contraband, the contraband in respect of which the inmate is convicted is forfeited to Her Majesty in right of Canada.

Expiration

(2) Where, on application made by an inmate in accordance with the regulations within three months after a forfeiture referred to in subsection (1), it is established to the satisfaction of the Commissioner or a member of the Service designated by the Commissioner that the forfeiture would cause undue hardship to the inmate, the Commissioner or that member shall, if possession of the object forfeited by the inmate would be lawful, cancel the forfeiture and order that the object be delivered to the inmate.

Idem

(3) Where, within three months after a forfeiture referred to in subsection (1), it is established to the satisfaction of the Commissioner or an officer of the Service designated by the Commissioner that a person other than the inmate has title to or an interest in an object forfeited and is innocent of any complicity in the events that resulted in the forfeiture, the Commissioner or that officer shall, if possession of the object forfeited by that person would be lawful, cancel the forfeiture and order that the object be delivered to that person.

Definition of "contraband"

(4) For the purposes of this section, "contraband" means anything that is in an inmate's possession in circumstances in which possession thereof is forbidden by any Act, regulation or Commissioner's directive, or by an order of general or specific application within the penitentiary in which the inmate is imprisoned. R.S., 1985, c. P-5, s. 35; R.S., 1985, c. 35 (2nd Supp.), s. 27.

ADMINISTRATION OF DECEASED INMATES' ESTATES

Service may administer estate

36. (1) Subject to and in accordance with any regulations made under subsection 37(1), the Service may, if the appropriate authority of the province in which the inmate was incarcerated does not do so, collect, administer and distribute the estate of a deceased inmate.

Definition of "estate of a deceased inmate"

(2) For the purposes of this section, "estate of a deceased inmate" means the following parts of the estate of an inmate who dies while serving a term of imprisonment in a penitentiary:

(a) any pay that, under the regulations, was due or otherwise payable to the inmate at the time of the inmate's death;

(b) any moneys standing to the inmate's credit at that time in any fund maintained or controlled by the Service; and

(c)   any personal belongings, including cash, found on the inmate or in the possession of the inmate at the time of death or that are in the care or custody of the Service at that time. 1976-77, c. 54, s. 43.

REGULATIONS AND RULES

Regulations

37.   (1)   The Governor in Council may make regulations

(a)   for the organization, training, discipline, efficiency, administration and good government of the Service;

(b)   for the custody, treatment, training, employment and discipline of inmates;

(c)   prescribing the compensation that may be paid pursuant to section 34, the terms and conditions in accordance with which the compensation is to be paid and the manner of its payment;

(c.1)   prescribing the manner in which an inmate applies for cancellation of a forfeiture of property under subsection 35(2);

(d)   defining the term "spouse" and the expression of "dependent child" for the purposes of section 34;

(e)   for the collection, administration and distribution of estates of deceased inmates;

(f)   providing for the appointment by the Governor in Council or by the Minister of a person to preside over a disciplinary court, prescribing the duties to be performed by that person and fixing that persons' remuneration; and

(g)   generally, for carrying into effect the purposes and provisions of this Act.

Punishment for contravention

(2) The Governor in Council may, in any regulations made under subsection (1) other than paragraph (b) thereof, provide for a fine not exceeding five hundred dollars or imprisonment for a term not exceeding six months, or both, to be imposed on summary conviction for the contravention of any such regulation.

Rules and orders of Commissioner

(3) Subject to this Act and any regulations made under subsection (1), the Commissioner may make rules, to be known as Commissioner's directives, for the organization, training, discipline, efficiency, administration and good government of the Service, and for the custody, treatment, training, employment and discipline of inmates and the good government of penitentiaries. R.S., 1985, c. P-5,s. 37; R.S., 19485, c. 35 (2nd Supp.), s. 28.

ANNUAL REPORT

Annual report

38. (3) The Minister shall, on or before January 31 next following the end of each fiscal year, or if parliament is not then sitting on any of the first five days next thereafter that either House of Parliament is sitting, submit to Parliament a report showing the operation of the Service for that fiscal year. R.S., c. P-6, s. 30.

RELATED PROVISIONS

R.S., 19485, c. 34 (2nd Supp.), s. 14:

Review of Act after three years

14. (1) Three years after the coming into force of this Act, a comprehensive review of the provisions and operation of sections 21.2 to 21.6 of the *Parole Act* and the schedule thereto, as enacted by sections 5 and 9 of this Act, and subsections 25(5) to (7) and section 26.1 of the *Penitentiary Act,* as enacted by subsection 10(2) and section 11 of this Act, shall be undertaken by such committee of the House of Commons or of both Houses of Parliament as may be designated or established by Parliament for that purpose.

Report to Parliament

(2) The committee referred to in subsection (1) shall, within one year after a review is undertaken pursuant to that subsection or within such further time as Parliament may authorize, submit to Parliament a report on the review, including a statement of any changes the committee recommends.

R.S., 1985, c. 35 (2nd Supp.), s. 15(2):

References to "Correctional Services of Canada"

"(2) Whenever the 'Canadian Penitentiary Service' is referred to in any Act of Parliament other than this Act or in any document, regulation or statutory instrument made thereunder, there shall in every case, unless the context otherwise requires, be substituted the 'Correctional Service of Canada.'"

R.S., 19485, c. 35 (2nd Supp.), s. 18(2):

References to members

"(2) Whenever, with respect to the Correctional Service of Canada, the word 'officer' or 'employee' occurs n any Act of Parliament other than this Act or any document, regulation or statutory instrument made thereunder, there shall in every case, unless the context otherwise requires, be substituted the word 'member.'"